ATLAS OF NEOTROPICAL LEPIDOPTERA

Volume 2

Checklist: Part 1

ATLAS OF

NEOTROPICAL LEPIDOPTERA

Checklist: Part 1

Micropterigoidea - Immoidea

Edited by

J. B. HEPPNER

Center for Arthropod Systematics
Florida State Collection of Arthropods
Gainesville, Florida

1984 **DR W. JUNK PUBLISHERS**
a member of the KLUWER ACADEMIC PUBLISHERS GROUP
THE HAGUE / BOSTON / LANCASTER

Distributors

for the United States and Canada: Kluwer Academic Publishers, 190 Old Derby Street, Hingham, MA 02043, USA
for the UK and Ireland: Kluwer Academic Publishers, MTP Press Limited, Falcon House, Queen Square, Lancaster LA1 1RN, England
for all other countries: Kluwer Academic Publishers Group, Distribution Center, P.O. Box 322, 3300 AH Dordrecht, The Netherlands

Library of Congress Cataloging in Publication Data

Main entry under title:

Atlas of neotropical lepidoptera.

Bibliography: v. 2, p.
Includes index.
Contents: -- 2. Checklist.
1. Lepidoptera--Latin America--Collected works.
2. Insects--Latin America--Collected works. I. Heppner, John B.
QL553.A1A85 1984 595.78'098 84-7139
ISBN 90-6193-038-3 (v. 2)

ISBN 90-6193-038-3 (this volume)
ISBN 90-6193-900-3 (series)

Cover design: Max Velthuijs

PRINTED IN THE NETHERLANDS

PATRONS

Bermuda
Costa Rica
Dominican Republic
Ecuador
Guyana
Jamaica
Netherlands
Peru
United States

Center for Arthropod Systematics, Florida State Collection of Arthropods, Gainesville, Florida
The Lepidoptera Research Foundation, Santa Barbara, California

To Latin American students and researchers
of Lepidoptera

CONTENTS

PREFACE

This checklist does not supercede any other because it is the first of its kind for the Neotropical region. It represents the culmination of several years preparation by the contributing authors and editor, following the initiation of the "Atlas" project in the autumn of 1978. The collaborators to the project continue their work, either for "Atlas" fascicles or for the more immediate completion of the remaining five parts of the checklist.

The goal of the project is to provide a foundation work on the Neotropical Lepidoptera, illustrated to the extent possible and encompassing a detailed text covering the basic data known for each described species (original citation, synonymy, diagnosis, range, flight period and hosts). New species will be described as each author finds useful, but the primary purpose is initially to document all the species already known. Efforts to document the Lepidoptera faunas of individual countries in Latin America are relatively illogical, since in most cases upwards of 80% of the entire Neotropical fauna is represented in each country having tropical forest and, thus, such works would have to deal with most of the continental fauna. The "Atlas" project covers the entire region to begin with but as a cooperative project among all New World nations.

The "Atlas of Neotropical Lepidoptera" will encompass 125 volumes in total, of which 115 volumes will comprise the illustrated descriptive text, one volume will be a general introductory volume on the Lepidoptera of the neotropics and an illustrated generic synopsis, six volumes will form the checklist, two volumes will be devoted to a butterfly and a moth bibliography for the Neotropical species, and a final volume will be a general index. The project is expected to take about 20 years to complete by present and future collaborating authors. Upon the completion of the entire project a revised checklist will be issued.

The present checklist part covers the first portion of the Microlepidoptera, the first 41 families presently known to occur in the Neotropical region, encompassed by the area from the Mexican-United States border south to Patagonia. The remaining Microlepidoptera, including Pyraloidea, will be treated in Part 2 of the checklist. Names are included that were published up to the end of 1983. New taxonomic data has been included, in terms of new synonymies and new combinations, to the extent known. Much of the new data or revised taxon arrangements reflect current and on-going research by the contributing authors, details of which will be published in revisionary treatments. Where appropriate, explanations are provided by running numbered notes, to be found at the end of the volume before the terminal index. The language of the "Atlas" is English so as to have a wider international audience for the publications.

Authorship is indicated for each family, both in the table of contents and in the main checklist. Individual authors are responsible for their parts in terms of omissions and errors of spelling and dates. The format of the checklist, the higher classification adopted, editing, preparation of introductory sections and the index, and preparation of the manuscripts for printing, are the responsibility of the editor.

To the contributing authors, to those nations already pledged to the financial support of the "Atlas" project, and to the publisher and other patrons, goes the great credit of a broad and long-range outlook for this endeavor.

Gainesville — J. B. Heppner

December 30, 1983 — Editor

ATLAS OF NEOTROPICAL LEPIDOPTERA: GENERAL PLAN

Volume 1	Introduction	
Vol. 2-7	Checklist: 6 pts.	
Vol. 8	Fasc. 1-14	Primitive moths
Vol. 9-10	Fasc. 15-20	Tineoidea
Vol. 11-14	Fasc. 21-29	Gelechioidea
Vol. 15	Fasc. 30-41	Copromorphoidea, Yponomeutoidea, Immoidea
Vol. 16-38	Fasc. 42-45	Pyraloidea, Pterophoroidea
Vol. 39-40	Fasc. 46-50	Sesioidea, Zygaenoidea
Vol. 41-42	Fasc. 51-55	Cossoidea, Castnioidea
Vol. 43	Fasc. 56	Tortricoidea
Vol. 44-71	Fasc. 57-61	Papilionoidea
Vol. 72-91	Fasc. 62-66	Geometroidea
Vol. 92-95	Fasc. 67-73	Bombycoidea
Vol. 96	Fasc. 74	Sphingoidea
Vol. 97-122	Fasc. 75-79	Noctuoidea
Vol. 123-124	Bibliography	
Vol. 125	General Index	

CHECKLIST OF NEOTROPICAL LEPIDOPTERA: GENERAL PLAN

Part 1	Micropterigoidea
	Heterobathmioidea
	Neopseustoidea
	Hepialoidea
	Nepticuloidea
	Incurvarioidea
	Tineoidea
	Gelechioidea
	Copromorphoidea
	Yponomeutoidea
	Immoidea
Part 2	Pyraloidea
	Pterophoroidea
	Sesioidea
	Zygaenoidea
	Cossoidea
	Castnioidea
	Tortricoidea
Part 3	Papilionoidea
Part 4	Geometroidea
	Bombycoidea
	Sphingoidea
Part 5	Noctuoidea
Part 6	General Index

CONTRIBUTING AUTHORS

Andrés O. Angulo	Universidad de Concepción, Concepcion, Chile (Noctuidae, in part)
Vitor O. Becker	Centro de Pesquisa Agropecuaria dos Cerrados, EMBRAPA, Planaltina, D.F., Brazil (Gelechioidea; Zygaenoidea, in part)
Keith S. Brown, Jr.	Universidade Estadual de Campinas, Campinas, São Paulo, Brazil (Nymphalidae, in part; Papilionidae, in part)
Richard L. Brown	Mississippi State University, Starkville, Mississippi, USA (Tortricidae, in part)
Curtis J. Callaghan	c/o Museo Nacional, Rio de Janeiro, Brazil (Lycaenidae, in part)
†Robert H. Carcasson	c/o Provincial Museum, Victoria, British Columbia, Canada (Sphingidae)
Mirna M. Casagrande	Universidade Federal do Paraná, Curitiba, Paraná, Brazil (Nymphalidae, in part)
Charles V. Covell, Jr.	University of Louisville, Louisville, Kentucky, USA (Geometridae, in part)
Donald R. Davis	Smithsonian Institution, Washington, D.C., USA (Primitive moths, in part; Tineoidea, in part; Epipyropidae)
Julian P. Donahue	Los Angeles County Museum of Natural History, Los Angeles, California, USA (Cossidae)
W. Donald Duckworth	Smithsonian Institution, Washington, D.C., USA (Sesiidae, in part)
Thomas D. Eichlin	California Department of Food and Agriculture, Sacramento, California, USA (Sesiidae, in part)
Douglas C. Ferguson	Systematic Entomology Laboratory, USDA, Washington, D.C., USA (Lymantriidae)
William D. Field	Smithsonian Institution, Washington, D.C., USA (Bibliography, in part)
John G. Franclemont	Cornell University, Ithaca, New York, USA (Bombycoidea, in part)
Hugh A. Freeman	Garland, Texas, USA (Hesperiidae, in part)
David F. Hardwick	Biosystematics Research Institute, Agriculture Canada, Ottawa, Ontario, Canada (Noctuidae: Heliothinae)
John B. Heppner	Center for Arthropod Systematics, Florida State Collection of Arthropods, FDACS, Gainesville, Florida, USA (Heliozelidae, Amphitheridae; Copromorphoidea; Yponomeutoidea; Immoidea; Sesioidea, in part; Tortricidae, in part; Hyblaeidae; Geometroidea, in part; Bombycoidea, in part; Noctuoidea, in part)

Charles L. Hogue	Los Angeles County Museum of Natural History, Los Angeles, California, USA (Noctuidae, in part)
Rienk de Jong	Rijksmuseum van Natuurlijke Historie, Leiden, Netherlands (Hesperiidae, in part)
Niels P. Kristensen	Zoologisk Museum, University of Copenhagen, Copenhagen, Denmark (Micropterigidae and Heterobathmiidae, in part; Neotheoridae)
J. Donald Lafontaine	Biosystematics Research Institute, Agriculture Canada, Ottawa, Ontario, Canada (Noctuidae, in part)
Gerardo Lamas M.	Museo de Historia Natural "Javier Prado," Lima, Peru (Papilionoidea, in part; bibliography, in part)
Claude Lemaire	c/o Museum National d'Histoire Naturelle, Paris, France (Saturniidae)
Olaf H. H. Mielke	Universidade Federal do Paraná, Curitiba, Paraná, Brazil (Hesperiidae, in part)
Jacqueline Y. Miller	Allyn Museum of Entomology, Sarasota, Florida, USA (Castniidae)
Lee D. Miller	Allyn Museum of Entomology, Sarasota, Florida, USA (Papilionoidea, in part)
Scott E. Miller	Museum of Comparative Zoology, Harvard University, Cambridge, Massachusetts, USA (Lyonetiidae; Gracillariidae, in part; Alucitidae; Pterophoridae; Dalceridae)
Eugene G. Munroe	Biosystematics Research Institute, Agriculture Canada, Ottawa, Ontario, Canada (Pyralidae, in part)
Stanley S. Nicolay	Virginia Beach, Virginia (Lycaenidae, in part)
E. Schmidt Nielsen	Division of Entomology, CSIRO, Canberra City, A.C.T., Australia (Micropterigidae, Heterobathmiidae, Hepialidae, and Palaeosetidae, in part)
Jerry A. Powell	University of California, Berkeley, California, USA (Tortricidae, in part)
Józef Razowski	Institute of Systematic Zoology, Polish Academy of Sciences, Krakow, Poland (Tortricidae, in part)
Gaden S. Robinson	British Museum (Natural History), London, England, U.K. (Hepialidae and Palaeosetidae, in part)
Jay C. Shaffer	George Mason University, Fairfax, Virginia, USA (Pyralidae, in part)
Gerhard Tarmann	Tiroler Landeskundliches Museum, Innsbruck, Austria (Zygaenidae)
Paul Thiaucourt	c/o Museum National d'Histoire Naturelle, Paris, France (Notodontidae)
Allan Watson	British Museum (Natural History), London, England, U.K. (Drepanidae; Arctiidae, in part)

Paul E. S. Whalley	British Museum (Natural History), London, England, U.K. (Thyrididae)
Chris Wilkinson	Vrije Universiteit, Amsterdam, Netherlands (Nepticulidae, in part)

INTRODUCTION

The present checklist part lists the described taxa of the Neotropical region for the first 41 families currently known to occur within the area chosen for this project. Inasmuch as northern Mexico, although faunistically a part of the Nearctic region, is not normally included in Nearctic publications, the "Neotropical" region for the "Atlas" project covers the area from the Mexican-United States border south to Patagonia, including the West Indies and all oceanic islands claimed by Latin American nations. Bermuda is also included. Hawaii (USA) and Easter Id. (Chile) are not included, since their faunas are Melanesian. The 4189 species listed herein involve all those having a type-locality within this geographical area, plus those northern species having distributions extending into Mexico. Also included are a few species which have been collected in the border area but for which no verified Mexican records are known; these species are indicated by an asterisk (*).

Checklist Format

The format of the checklist is rather straightforward and follows standard entomological checklist style in listing valid species under valid generic names, with synonyms listed by date under each category. Author date follows each name, with parentheses added to indicate names not in their original generic combinations. The type-locality is noted for each name by nation only, except Brazil, Mexico, and the United States, where the states are also indicated (abbreviations used are listed on the pages covering abbreviations below). Localities noted from derived information are in brackets. All names adopted as current valid names for genera, species and subspecies are noted in boldface type. All synonyms are listed in Roman type face (this is likewise followed in the indeces). A synopsis of all included genera, including synonyms, is presented prior to the actual checklist.

Classification

Higher categories follow the arrangement proposed in a new book on the classification of Lepidoptera families (Heppner, 1984b). The new arrangement follows recent classifications, like that of Common (1970), but changes have been adopted following more recent studies. Synonyms of higher category names are not listed. All current names below the family level are listed in a phylogenetic arrangement, beginning with the taxa thought to be most primitive within each category to those thought to be most advanced. Exceptions to a phylogenetic arrangement involve only those groups which are so little known that an alphabetical arrangement is the only rational approach at this time. In Tineidae an alphabetical arrangement has also been chosen by the author, since the subfamilies require extensive revision.

Nomenclature

All described Neotropical species known to the authors for the first 41 families are included in the checklist, including all infraspecific names, emendations and misspellings. Rejected names are also listed in synonymy, as are major misidentifications often causing confusion in the literature. Various abbreviations are used for these and other situations, as noted in the list of descriptors below. Dates of authorship of names are as accurate as is known and those of certain older works follow the dates noted in Heppner (1982c) and various opinions of the Commission on Zoological Nomenclature. Date ranges and dates derived from other information are in brackets. The most recent date in a date range is always taken as the date of record in cases of priority. Original spellings of taxa are used thoughout, except for the deletion of diacritic marks which are not allowed by Article 27 of the Zoological Code. Terminal notes give information on particular aspects of certain names and are consecutively numbered throughout the checklist.

New synonyms and new combinations are clearly indicated by boldface type for the appropriate abbreviations (e.g., **n. syn.**). A few names are transferred in the checklist from one family to another and these names are listed below. One family, the Palaeosetidae, is listed for the first time for the Neotropical fauna.

Indexing

All families listed are given a reference number, as also indicated in the table of contents. This reference number is also used in the index to allow species names to be found in the text. Species are numbered separately for each family to allow easier future expansion of numbers. Each specific name is indexed in the terminal index by reference to its family number and its species number within that family; e.g., 29-200 refers to species 200 in family 29, which is Gelechiidae. Once readers become familiar with families of interest to them then later reference to index entries will easily identify specific names of these families merely by the family number. These number codes will also later be used in "Atlas" fascicles, other than for species that are transferred because of future revisionary studies. The generic index has the names indexed by page number.

ACKNOWLEDGMENTS

Many persons have been involved in the initial planning of the project, not least of which are the authors collaborating on the "Atlas," but W. D. Field (Smithsonian Institution, Washington) and E. L. Todd (Systematic Entomology Laboratory, USDA, Washington), deserve special mention for their encouragement. Illness has, unfortunately, prevented Dr. Todd from making his planned contributions in the Noctuidae fascicle of the "Atlas."

Mr. Wil R. Peters, the publisher for Dr. W. Junk Publishers B.V., The Hague, Netherlands, has been instrumental in fostering the "Atlas" project. His patience during these beginning years since 1978 has been crucial to the continued success of the project.

The nations thus far in support of the "Atlas" project are especially thanked for their foresight. The "Atlas," as a cooperative project, is the most logical and least expensive way for each Latin American nation to provide the basic documentation of native Lepidoptera, as noted previously, due to the generally widespread distributions of many Neotropical species. For the Netherlands and their New World interests, the Stichting voor Wetenschappelijk Orderzoek van de Troppen, The Hague, should be especially thanked here for providing already since 1981 the basic funding which in particular allowed the indexing of this first checklist volume to be completed on time (grant W87-173).

The Lepidoptera Research Foundation, Santa Barbara, California, has recently agreed to handle financial matters for the "Atlas" project by receiving donations from supporting nations, other organizations, and individuals.

A final word on one of the collaborators. Dr. Robert H. Carcasson, who passed away in September, 1982, was completing the checklist manuscript for the Neotropical Sphingidae as one of his last works, to be published in Part 4 of the Neotropical checklist. This work will be published with Dr. Carcasson as author as he expected.

TRANSFERRED TAXA

Mallomus Blanchard, 1854 - from Hepialidae to Geometridae
Mallomus ciliatus Blanchard, 1854 - from Hepialidae to Geometridae (see Checklist, Part 4)

"Gaea" lilloi Köhler, 1941 - from Sesiidae to Gelechiidae

Pexicnemidia Möschler, 1890 - from Tineidae to Yponomeutidae
Pexicnemidia mirella Möschler, 1890 - from Tineidae to Yponomeutidae

"Dalaca" perkeo Pfitzner, 1914 - from Hepialidae to Cossidae (see Checklist, Part 2)

Syncerastis Meyrick, 1931 - from Tineidae to Yponomeutidae
Syncerastis ptisanopa Meyrick, 1931 - from Tineidae to Yponomeutidae

Sericostola Meyrick, 1927 - from Yponomeutidae to Glyphipterigidae
Sericostola rhodanopa Meyrick, 1927 - from Yponomeutidae to Glyphipterigidae

Elachista rubella Blanchard, 1852 - from Elachistidae to Heliodinidae (**Heliodines**)

ABBREVIATIONS

Descriptors

ab.	aberration
auct.	of authors (= misdetermination)
aut. f.	autumn form
emend.	emendation (incorrect emendation)
extralim.	extralimital (not occurring in the Neotropical region; usually involving an extralimital species having Neotropical subspecies)
f.	form
gynand.	gynandromorph
hyb.	hybrid
incorr. spell.	incorrect spelling (use of diacritic marks not allowed by ICZN)
misid.	misidentified
mispl.	misplaced (generic placement uncertain or incorrect)
missp.	misspelling
n. comb.	new combination
n. syn.	new synonymy
nom. dub.	nomen dubium (= identity uncertain)
nom. nud.	nomen nudum (= undescribed name)
nom. oblit.	nomen oblitum (= forgotten name; unused senior synonym)
preocc.	preoccupied name
redesc.	redescribed name (a second original description)
repl. name	replacement name (for a preoccupied name)
rev. stat.	revised status
[sic]	original misspelling (= lapsus calami)
spr. f.	spring form
sum. f.	summer form
suppr.	suppressed (or in a rejected work according to ICZN)
unassoc.	unassociated female or male
unavail.	unavailable name (by ICZN article)
uncert. stat.	status uncertain
var.	variety
wint. f.	winter form
*	prefix for extralimital taxa (also for species from the United States border area not yet collected in northern Mexico but which are presumed to occur in Mexico as well)
/	= "and" (used to separate two or more localities where a lectotype has not yet been designated to restrict the type-locality)
ICZN	International Commission on Zoological Nomenclature
teste	= verified

Abbreviations for States of Brazil

Acre	Ac
Alagoas	Al
Amapa	Ap
Amazonas	Am
Bahia	Ba
Ceara	Ce
Distrito Federal	DF
Espirito Santo	ES
Goiás	Go
Maranhao	Ma
Mato Grosso	MT
Minas Gerais	MG
Pará	Pa
Paraiba	Pb
Paraná	Pr
Pernambuco	Pe
Piaui	Pi
Rio de Janeiro	RJ
Rio Grande do Norte	RN
Rio Grande do Sul	RS
Rondonia	Rd
Roraima	Rr
Santa Catarina	SC
Sao Paulo	SP
Sergipe	Se

Abbreviations for States of Mexico

Aguascalientes	Agu	Morelos	Mor
Baja California Norte	BCN	Nayarit	Nay
Baja California Sur	BCS	Nuevo Leon	NL
Campeche	Cam	Oaxaca	Oax
Chiapas	Chia	Puebla	Pue
Chihuahua	Chih	Queretaro	Que
Coahuila	Coa	Quintana Roo	QR
Colima	Col	San Luis Potosi	SLP
Distrito Federal	DF	Sinaloa	Sin
Durango	Dur	Sonora	Son
Guanajuato	Gua	Tabasco	Tab
Guerrero	Gue	Tamaulipas	Tam
Hidalgo	Hid	Tlaxcala	Tla
Jalisco	Jal	Veracruz	Ver
Mexico	Mex	Yucatan	Yuc
Michoacan	Mic	Zacatecas	Zac

Abbreviations for States of the United States

Alabama	Al	Montana	Mt
Alaska	Ak	Nebraska	Nb
Arizona	Az	Nevada	Nv
Arkansas	Ar	New Hampshire	NH
California	Ca	New Jersey	NJ
Colorado	Co	New Mexico	NM
Connecticut	Ct	New York	NY
Delaware	De	North Carolina	NC
District of Columbia	DC	North Dakota	ND
Florida	Fl	Ohio	Oh
Georgia	Ga	Oklahoma	Ok
Hawaii	Hi	Oregon	Or
Idaho	Id	Pennsylvania	Pa
Illinois	Il	Rhode Island	RI
Indiana	In	South Carolina	SC
Iowa	Ia	South Dakota	SD
Kansas	Ks	Tennessee	Tn
Kentucky	Ky	Texas	Tx
Louisiana	La	Utah	Ut
Maine	Me	Vermont	Vt
Maryland	Md	Virginia	Va
Massachusetts	Ma	Washington	Wa
Michigan	Mi	West Virginia	WV
Minnesota	Mn	Wisconsin	Wi
Mississippi	Ms	Wyoming	Wy
Missouri	Mo		

OUTLINE OF CLASSIFICATION OF THE LEPIDOPTERA

The list of families, subfamilies, and associated higher categories, presented here shows in outline form the classification adopted for the "Atlas of Neotropical Lepidoptera." The classification follows that of a book recently completed by the editor (Heppner, 1984b). This classification is similar to that of Common (1970) but includes the most recent changes based on current understandings of various specialists (e. g., Brock, 1971; Common, 1975; Davis, 1975b, 1978a, 1978b; Davis and Nielsen, 1980; Heppner, 1977, 1982a, 1982b; Hodges, 1978; Hodges, et al., 1983; Kristensen, 1978a, 1978b; Kristensen and Nielsen, 1979, 1980; Nielsen and Davis, 1981; Powell, 1980; Razowski, 1974; Zagulajev, 1973; Zimmerman, 1978), as well as such innovations as the placement of Amphitheridae in Tineoidea, the placement of Pyraloidea after Immoidea, and the rearrangement of families within the Cossoidea and Zygaenoidea. Specific documentation for these latter changes will be found in Heppner (1984b). Refinements may be made in the future as new information becomes available to possibly alter our concepts of Lepidoptera phylogeny. The families known to occur in the Neotropical region are numbered in the checklist (as also in the Table of Contents) and these numbers are used in the index to refer species names to their position in the checklist. In the outline the unnumbered families are extralimital and not known to occur in the Neotropical region. The number following each Neotropical family for the first 41 families indicates the species total listed in the checklist for that family; the numbers for families 42 to 79 involve the estimated numbers of described species for the Neotropical region as noted in Heppner (1984b) (some species totals for families do not correspond to the number indicated in Heppner (1984b) due to the inclusion herein of Nearctic species that overlap the border region).

Under Tineidae recognized subfamilies are listed but since extensive revision is needed to redefine the limits of these subfamilies, possibly with major synonymies, the genera and species have been listed in the checklist entirely alphabetically for this family.

Some major recent literature references that revise the Neotropical fauna for the families covered in this first part of the checklist, or otherwise form the basis of current taxonomic decisions by the authors, are as follows:

Micropterigidae (Kristensen and Nielsen, 1979, 1982)
Heterobathmiidae (Kristensen and Nielsen, 1979, 1983)
Neopseustidae (Davis, 1975b; Davis and Nielsen, 1980)
Neotheoridae (Kristensen, 1978a)
Hepialidae Robinson, 1977; Robinson and Nielsen, 1983)
Palaeosetidae (Robinson and Nielsen, 1983)
Nepticulidae (Newton and Wilkinson, 1982; Scoble, 1982, 1983; Wilkinson and Newton, 1981)
Opostegidae (Davis, 1984; Eyer, 1964)
Tischeriidae (Braun, 1972; Hering, 1926)
Incurvariidae (Nielsen and Davis, 1981)
Cecidosidae (Becker, 1977)
Adelidae (Nielsen, 1980; Pastrana, 1961; Razowski and Wojtusiak, 1978)
Prodoxidae (Davis, 1967)
Heliozelidae (Wojtusiak, 1978)
Tineidae (Davis, 1978b; Dietz, 1905; Hasbrouck, 1964; Petersen, 1957-58; Robinson, 1976, 1978; Zagulajev, 1960-79, 1973)
Psychidae (Davis, 1964, 1975a; Vázquez, 1941-[44])
Arrhenophanidae (Bradley, 1951)
Amphitheridae (Moriuti, 1978)
Lyonetiidae (Braun, 1963; Kuroko, 1964)
Gracillariidae (Ely, 1918; Opler and Davis, 1981; Vári, 1961)
Oecophoridae (Becker, 1982; Clarke, 1971, 1978; Duckworth, 1969, 1970, 1971; Hodges, 1974; Kasy, 1973; Powell, 1973)
Elachistidae (Braun, 1948; Traugott-Olsen and Nielsen, 1977)
Blastobasidae (Dietz, 1910; Gozmány, 1963)
Coleophoridae (Hodges, 1966b; Sattler and Tremewan, 1974, 1978; Toll, 1962)
Momphidae (Hodges, 1966b; Riedl, 1969)
Agonoxenidae (Clarke, 1962)
Cosmopterigidae (Hodges, 1978)
Scythrididae (Jäckh, 1977-78; Passerin d'Entrèves, 1977)
Gelechiidae (Busck, 1903b, 1939; Hodges, 1966a; Povolný, 1967a, 1967b; Sattler, 1960, 1973)
Copromorphidae (Clarke, 1955; Heppner, 1984a, 1984c)
Alucitidae (Buszko, 1977)
Carposinidae (Davis, 1969)
Epermeniidae (Gaedike, 1977)
Glyphipterigidae (Diakonoff, 1984; Heppner, 1981a, 1981b, 1982a, 1984d; Toll, 1956)
Plutellidae (Busck, 1903a; Moriuti, 1977)
Yponomeutidae (Moriuti, 1977)
Argyresthiidae (Moriuti, 1977)
Douglasiidae (Gaedike, 1974)
Acrolepiidae (Davis, 1980; Gaedike, 1970)
Heliodinidae (Gaedike, 1967; Meyrick, 1914)
Immidae (Heppner, 1982d)

Classification Outline

LEPIDOPTERA

Suborder Zeugloptera

Micropterigoidea

1. **Micropterigidae** (2)

Suborder Aglossata

Agathiphagoidea

* Agathiphagidae

Suborder Heterobathmiina

Heterobathmioidea

2. **Heterobathmiidae** (2)

Suborder Glossata

Infraorder Dacnonypha

Eriocranioidea

- * Eriocraniidae
- * Acanthopteroctetidae
- * Lophocoronidae

Infraorder Neopseustina

Neopseustoidea

3. **Neopseustidae** (4)

Infraorder Exoporia

Mnesarchaeoidea

- * Mnesarchaeidae

Hepialoidea

4. **Neotheoridae** (1)
* Anomosetidae
* Prototheoridae
5. **Hepialidae** (131)
6. **Palaeosetidae** (1)

Infraorder Heteroneura

Division Monotrysia

Nepticuloidea

7. **Nepticulidae** (13)
Pectinivalvinae
Nepticulinae
8. **Opostegidae** (16)
9. **Tischeriidae** (8)

Incurvarioidea

10. **Incurvariidae** (4)
11. **Cecidosidae** (4)
12. **Adelidae** (17)
13. **Prodoxidae** (6)
14. **Heliozelidae** (15)

Division Ditrysia

Section Tineina

Tineoidea

15. **Tineidae** (456)
Nemapogoninae
Scardiinae
Euplocaminae
Tinissinae
Meessiinae
Teichobiinae
Dryadaulinae
Acrolophinae
Setomorphinae
Phthoropoeinae
Myrmecozelinae
Siloscinae
Tineinae
Erechthiinae
Hieroxestinae
* Eriocottidae
Eriocottinae
Compsocteninae
16. **Psychidae** (73)
Penestoglossinae
Taleporiinae
Psychinae
Oiketicinae
* Pseudarbelidae
17. **Arrhenophanidae** (7)
* Ochsenheimeriidae
18 **Amphitheridae** (1)
19. **Lyonetiidae** (19)
Bedelliinae
Lyonetiinae
Cemiostominae
Bucculatriginae
20. **Gracillariidae** (147)
Gracillariinae
Lithocolletinae
Phyllocnistinae

Gelechioidea

21. **Oecophoridae** (1733)
Depressariinae
Ethmiinae
Peleopodinae
Autostichinae
Xyloryctinae
Stenomatinae
Oecophorinae
Hypertrophinae
Chimabachinae
Deuterogoniinae
22. **Elachistidae** (7)
Coelopoetinae
Elachistinae
* Pterolonchidae
23. **Blastobasidae** (114)
Symmocinae
Blastobasinae
24. **Coleophoridae** (32)
Coleophorinae
Batrachedrinae
25. **Momphidae** (29)
26. **Agonoxenidae** (48)
Agonoxeninae
Blastodacninae
27. **Cosmopterigidae** (110)
Antequerinae
Cosmopteriginae
Chrysopeleiinae
28. **Scythrididae** (12)
29. **Gelechiidae** (835)
Anomologinae
Gelechiinae
Anacampsinae
Chelariinae
Dichomeridinae

Lecithocerinae
Physoptilinae

Copromorphoidea

30. **Copromorphidae** (12)
31. **Alucitidae** (24)
32. **Carposinidae** (7)
33. **Epermeniidae** (3)
    Epermeniinae
    Ochromolopinae
34. **Glyphipterigidae** (75)

Yponomeutoidea

35. **Plutellidae** (27)
    Orthotaeliinae
    Galacticinae
    Plutellinae
    Scythropinae
    Praydinae
36. **Yponomeutidae** (90)
    Attevinae
    Saridoscelinae
    Yponomeutinae
    Cedestinae
37. **Argyresthiidae** (7)
38. **Douglasiidae** (1)
39. **Acrolepiidae** (21)
40. **Heliodinidae** (39)
    Schreckensteiniinae
    Heliodininae

Immoidea

41. **Immidae** (36)

Pyraloidea

42. **Hyblaeidae** [2]
43. **Thyrididae** [183]
    Argyrotypinae
    Thyridinae
    Siculodinae
    Striglininae
* Lathrotelidae
44. **Pyralidae** [3350]
    Midilinae
    Scopariinae
    Nymphulinae
    Odontiinae
    Glaphyriinae
    Evergestinae
    Pyraustinae
    Schoenobiinae
    Cybalomiinae
    Crambinae
    Pyralinae
    Chrysauginae
    Epipaschiinae
    Galleriinae
    Phycitinae
    Peoriinae

Pterophoroidea

* Tineodidae
* Oxychirotidae
45. **Pterophoridae** [135]
    Agdistinae
    Platyptiliinae
    Pterophorinae

Sesioidea

46. **Brachodidae** [13]
    Brachodinae
    Phycodinae
47. **Sesiidae** [193]
    Tinthiinae
    Paranthreninae
    Sesiinae
48. **Choreutidae** [91]
    Millieriinae
    Brenthiinae
    Choreutinae

Zygaenoidea

* Heterogynidae
49. **Zygaenidae** [131]
    Zygaeninae
    Phaudinae
    Charideinae
    Chalcosiinae
    Anomoeotinae
    Himantopterinae
    Procridinae
* Somabrachyidae
50. **Megalopygidae** [251]

Section Cossina

Cossoidea

51. **Cossidae** [166]
    Pseudocossinae
    Hypoptinae
    Cossulinae
    Cossinae
    Zeuzerinae
* Dudgeoneidae
* Metarbelidae
* Cyclotornidae
52. **Epipyropidae** [4]
53. **Dalceridae** [80]
    Acraginae
    Dalcerinae
54. **Limacodidae** [260]
* Chrysopolomidae
    Ectropinae
    Chrysopolominae

Castnioidea

55. **Castniidae** [160]

Synemoninae
Neocastniinae
Castniinae

Tortricoidea

56. **Tortricidae** [676]
Chlidanotinae
Ceracinae
Tortricinae
Cochylinae
Olethreutinae

Papilionoidea

57. **Hesperiidae** [1786]
Megathyminae
Coeliadinae
Pyrrhopyginae
Pyrginae
Trapezitinae
Heteropterinae
Hesperiinae
58. **Papilionidae** [119]
Baroniinae
Parnassiinae
Papilioninae
59. **Pieridae** [275]
Pseudopontiinae
Dismorphiinae
Pierinae
Coliadinae
60. **Lycaenidae** [2575]
Poritiinae
Miletinae
Curetinae
Lycaeninae
Riodininae
61. **Nymphalidae** [2798]
Danainae
Ithomiinae
Acraeinae
Heliconiinae
Nymphalinae
Calinaginae
Libytheinae
Apaturinae
Amathusiinae
Morphinae
Brassolinae
Satyrinae

Geometroidea

Series Calliduliformes

* Callidulidae
* Pterothysanidae
Pterothysaninae
Hibrildinae
* Ratardidae

Series Uraniformes

62. **Sematuridae** [35]
63. **Uraniidae** [6]
Microniinae
Uraniinae
* Epicopeiidae
* Apoprogonidae
64. **Epiplemidae** [230]

Series Geometriformes

65. **Geometridae** [7804]
Archiearinae
Oenochrominae
Ennominae
Geometrinae
Sterrhinae
Larentiinae

Series Drepaniformes

* Axiidae
66. **Thyatiridae** [5]
* Cyclidiidae
* Drepanidae
Drepaninae
Oretinae
Nidarinae

Bombycoidea

Series Bombyciformes

* Carthaeidae
67. **Eupterotidae** [3]
Striphnopteryginae
Janinae
Eupterotinae
68. **Apatelodidae** [245]
Apatelodinae
Epiinae
* Bombycidae
69. **Mimallonidae** [250]
* Anthelidae
Munychryiinae
Anthelinae
70. **Lasiocampidae** [707]
Chondrosteginae
Chionopsychinae
Archaeopachinae
Gonometinae
Macromphaliinae
Gastropachinae
Lasiocampinae

Series Saturniiformes

* Endromidae
* Lemoniidae
Lemoniinae
Sabaliinae
* Brahmaeidae
71. **Oxytenidae** [60]
72. **Cercophanidae** [30]
Cercophaninae
Janiodinae
73. **Saturniidae** [809]

Arsenurinae
Ceratocampinae
Hemileucinae
Agliinae
Ludiinae
Salassinae
Saturniinae

Sphingoidea

74. **Sphingidae** [397]
Sphinginae
Macroglossinae

Noctuoidea

75. **Notodontidae** [1650]
Notodontinae
Dilobinae
Thaumetopoeinae
76. **Dioptidae** [500]
* Thyretidae
77. **Lymantriidae** [180]
78. **Arctiidae** [6300]
Aganainae
Pericopinae
Lithosiinae
Arctiinae
Ctenuchinae
79. **Noctuidae** [8516]
Hypenodinae
Hypeninae
Herminiinae
Rivulinae
Ophiderinae
Catocalinae
Plusiinae
Euteliinae
Stictopterinae
Sarrothripinae
Chloephorinae
Nolinae
Acontiinae
Pantheinae
Acronictinae
Agaristinae
Cocytiinae
Amphipyrinae
Cuculliinae
Hadeninae
Noctuinae
Heliothinae

BIBLIOGRAPHY

Becker, V. O.
1977. The taxonomic position of the Cecidosidae Brèthes (Lepidoptera). Polski Pismo Ent., 47: 79-86.
1982. Stenomine moths of the Neotropical genus *Timocratica* (Oecophoridae). Bull. Br. Mus. (Nat. Hist.), Ent., 45:211-306.
Bradley, J. D.
1951. Notes on the family Arrhenophanidae (Lepidoptera: Heteroneura), with special reference to the morphology of the genitalia, and descriptions of one new genus and two new species. Ent. (London), 84:178-185.
Braun, A. F.
1948. Elachistidae of North America (Microlepidoptera). Mem. Amer. Ent. Soc., 13:1-110, 26 pl.
1963. The genus *Bucculatrix* in America north of Mexico (Microlepidoptera). Mem. Amer. Ent. Soc., 18:1-208, 45 pl.
1972. Tischeriidae of America north of Mexico (Microlepidoptera). Mem. Amer. Ent. Soc., 28:1-148.
Brock, J. P.
1971. A contribution towards an understanding of the morphology and phylogeny of the ditrysian Lepidoptera. J. Nat. Hist., 5:29-102.
Busck, A.
1903a. Notes on the *Cerostoma* group of Yponomeutidae, with description of new North American species. J. New York Ent. Soc., 9:45-59.
1903b. A revision of the American moths of the family Gelechiidae, with descriptions of new species. Proc. U. S. Natl. Mus., 25:767-938, pl. 28-32.
1939. Restriction of the genus *Gelechia* (Lepidoptera: Gelechiidae), with descriptions of new genera. Proc. U. S. Natl. Mus., 86:563-593, pl. 59-71.
Buszko, J.
1977. Alucitidae. In, Klucze do oznaczania owadów Polski, 27 (Lepidoptera), 38:1-18. Warsaw. [In Polish].
Clarke, J. F. G.
1955. Copromorphidae. In, Catalogue of the type specimens of Microlepidoptera in the British Museum (Natural History) described by Edward Meyrick, 2:509-531. London: Br. Mus. (Nat. Hist.).
1962. The genus *Homoeoprepes* Walsingham (Lepidoptera: Blastodacnidae). Neotropical Microlepidoptera, I and II. Proc. U. S. Natl. Mus., 113:373-382.
1971. Neotropical Microlepidoptera, XIX: notes on and new species of Oecophoridae (Lepidoptera). Smithson. Contr. Zool., 95:1-39.
1978. Neotropical Microlepidoptera, XXI: new genera and species of Oecophoridae from Chile. Smithson. Contr. Zool., 273:1-79.
Common, I. F. B.
1970. Lepidoptera (moths and butterflies). In, The insects of Australia: a textbook for students and research workers. Pp. 765-866. Canberra: Melbourne Univ. Press.
1975. Evolution and classification of the Lepidoptera. Ann. Rev. Ent., 20:183-203.
Davis, D. R.
1964. Bagworm moths of the Western Hemisphere (Lepidoptera: Psychidae). Bull. U. S. Natl. Mus., 244:1-233.
1967. A revision of the moths of the subfamily Prodoxinae (Lepidoptera: Incurvariidae). Bull. U. S. Natl. Mus., 255:1-170.
1969. A revision of the American moths of the family Carposinidae (Lepidoptera: Carposinoidea). Bull. U. S. Natl. Mus., 289:1-105.
1975a. A review of the West Indian moths of the family Psychidae with descriptions of new taxa and immature stages. Smithson. Contr. Zool., 188:1-66.
1975b. Systematics and zoogeography of the family Neopseustidae with the proposal of a new superfamily (Lepidoptera: Neopseustoidea). Smithson. Contr. Zool., 210:1-45.
1978a. A revision of the North American moths of the superfamily Eriocranioidea with the proposal of a new family, Acanthopteroctetidae (Lepidoptera). Smithson. Contr. Zool., 251:1-131.
1978b. The North American moths of the genera *Phaeoses, Opogona,* and *Oinophila,* with a discussion of their supergeneric affinities (Lepidoptera: Tineidae). Smithson. Contr. Zool., 282:1-39.
1980. A redescription of *"Micropteryx" selectella* Walker with a discussion concerning its family affinities (Acrolepiidae). J. Lepid. Soc., 32 187-190.
1984. A preliminary review of the family Opostegidae, with descriptions of new taxa and immature stages. Smithson. Contr. Zool., [in press].
Davis, D. R., and E. S. Nielsen
1980. Description of a new genus and two new species of Neopseustidae from South America, with discussion of phylogeny and biological observations (Lepidoptera: Neopseustoidea). Steenstrupia (Copenhagen), 6:253-289.
Diakonoff, A. N.
1984. Glyphipterigidae, sensu lato. In, H. G. Amsel, et al. (eds.), Microlepidoptera Palearctica, volume 7. Vienna: G. Fromme. [In press].
Dietz, W. G.
1905. Revision of the genera and species of the tineid subfamilies Amydriinae and Tineinae inhabiting North America. Trans. Amer. Ent. Soc., 31:1-95, pl. 1-6.
1910. Revision of the Blastobasidae of North America. Trans. Amer. Ent. Soc., 36:1-72, pl. 1-4.
Duckworth, W. D.
1969. Bredin-Archbold-Smithsonian biological survey of Dominica: West Indian Stenomidae (Lepidoptera: Gelechioidea). Smithson. Contr. Zool., 4:1-21.

1970. Neotropical Microlepidoptera, XVIII: revision of the genus *Peleopoda* (Lepidoptera: Oecophoridae). Smithson. Contr. Zool., 48:1-30.

1971. Neotropical Microlepidoptera, XX: revision of the genus *Setiostoma* (Lepidoptera: Stenomidae). Smithson. Contr. Zool., 106:1-45.

Ely, C. R.

1918. A revision of the North American Gracilariidae [sic] from the standpoint of venation. Proc. Ent. Soc. Wash., 19:29-77. (1917)

Eyer, J. R.

1964. A pictorial key to the North American moths of the family Opostegidae. J. Lepid. Soc., 17: 237-242.

Gaedike, R.

1967. Zur systematischen Stellung einiger Gattungen der Heliodinidae Schreckensteiniidae sowie Revision der paläarktischen Arten der Gattung *Pancalia* Curtis, 1830 (Lepidoptera). Beitr. Ent. (Berlin), 17:363-374.

1970. Revision der paläarktischen Acrolepiidae (Lepidoptera). Ent. Abh. (Dresden), 38:1-54.

1974. Revision der paläarktischen Douglasiidae (Lepidoptera). Acta Faun. Ent. Mus. Natl. Prague, 15:79-101.

1977. Revision der nearktischen und neotropischen Epermeniidae (Lepidoptera). Beitr. Ent. (Berlin), 27:301-312.

Gozmány, L. A.

1963. The family Symmocidae and the descriptions of new taxa mainly from the Near East (Lepidoptera). Acta Zool. Acad. Sci. Hung. (Budapest), 9:67-134.

Hasbrouck, F. F.

1964. Moths of the family Acrolophidae in America north of Mexico (Microlepidoptera). Proc. U. S. Natl. Mus., 114:487-706.

Heppner, J. B.

1977. The status of the Glyphipterigidae and a reassessment of relationships in yponomeutoid families and ditrysian superfamilies. J. Lepid. Soc., 31:124-134.

1981a. Revision of the new genus *Diploschizia* (Lepidoptera: Glyphipterigidae) for North America. Fla. Ent., 64:309-336.

1981b. *Neomachlotica*, a new genus of Glyphipterigidae (Lepidoptera). Proc. Ent. Soc. Wash., 83: 479-488.

1982a. Synopsis of the Glyphipterigidae (Lepidoptera: Copromorphoidea). Proc. Ent. Soc. Wash., 84:38-66.

1982b. A world catalog of genera associated with the Glyphipterigidae auctorum (Lepidoptera). J. New York Ent. Soc., 89:220-294.

1982c. Dates of selected Lepidoptera literature for the Western Hemisphere fauna. J. Lepid. Soc., 36:87-111.

1982d. Review of the family Immidae with a world checklist (Lepidoptera: Immoidea). Entomography (Sacramento), 1:257-279.

1984a. Revision of the Oriental and Nearctic genus *Ellabella* (Lepidoptera: Copromorphoidea). J. Res. Lepid. (Santa Barbara), 22: [in press].

1984b. Lepidoptera family classification. A guide to the higher categories, world diversity and literature resources of the butterflies and moths. Medford, New Jersey: Plexus Publ. [In press].

1984c. Revision of the genus *Lotisma* (Lepidoptera: Copromorphidae). J. Res. Lepid., 23: [in press].

1984d. Sedge moths of North America (Lepidoptera: Glyphipterigidae). Gainesville: Flora and Fauna Press. (ca. 185 pp.)

Hering, O.

1926. Die Blattminierer-Gattung *Tischeria* in ihren paläarktischen Arten. Ent. Jahrb. (Leipzig), 35: 99-106, 1 pl.

Hodges, R. W.

1966a. Revision of Nearctic Gelechiidae, I. The *Lita* group (Lepidoptera: Gelechioidea). Proc. U. S. Natl. Mus., 119:1-66, pl. 1-31.

1966b. Review of the New World species of *Batrachedra* with descriptions of three new genera (Lepidoptera: Gelechioidea). Trans. Amer. Ent. Soc., 92:585-651.

1974. Gelechioidea. Oecophoridae. In, R. B. Dominick, et al. (eds.), The moths of America north of Mexico. Fasc. 6.2. London: E. W. Classey. 142 pp., 8 pl.

1978. Gelechioidea. Cosmopterigidae. In, R. B. Dominick, et al. (eds.), The moths of America north of Mexico. Fasc. 6.1. London: E. W. Classey. 166 pp., 6 pl.

Hodges, R. W., et al. (eds.).

1983. Check list of the Lepidoptera of America north of Mexico. London: E. W. Classey. 284 pp.

Jäckh, E.

1977-78. Bearbeitung der Gattung *Scythris* Hübner (Lepidoptera, Scythrididae). Dtsch. Ent. Zeit. (Berlin), 24:261-271 (1977); 25:71-89 (1978).

Kasy, F.

1973. Beitrag zur Kenntnis der Familie Stathmopodidae Meyrick (Lepidoptera, Gelechioidea). Tijds. Ent. (Amsterdam), 116:227-299.

Kuroko, H.

1964. Revisional studies on the family Lyonetiidae of Japan (Lepidoptera). Esakia (Fukuoka), 4:1-61, pl. 1-17.

Kristensen, N. P.

1978a. A new familia of Hepialoidea from South America, with remarks on the phylogeny of the subordo Exoporia (Lepidoptera). Ent. Germanica (Stuttgart), 4:272-294.

1978b. Observations on *Anomoses hylecoetes* (Anomosetidae), with a key to hepialoid families (Insecta: Lepidoptera). Steenstrupia (Copenhagen), 5:1-19.

Kristensen, N. P., and E. S. Nielsen

1980. The ventral diaphragm of primitive (non-ditrysian) Lepidoptera. A morphological and phylogenetic study. Zeit. Zool. Syst. Evolutionsfor. (Frankfurt), 18:123-146.

1982. South American micropterigid moths: two new genera of the *Sabatinca*-group (Lepidoptera: Micropterigidae). Ent. Scand., 13:513-529.

1983. The *Heterobathmia* life history elucidated: immature stages contradict assignment to suborder Zeugloptera (Insecta: Lepidoptera). Zeit. Zool. Syst. Evolutionsfor. (Frankfurt), 21:101-124.

Meyrick, E.

1914. Lepidoptera Heterocera. Fam. Heliodinidae. Genera Insectorum (Brussels), 165:1-29, 2 pl.

Moriuti, S.
1977. Yponomeutoidea s. lat. (Insecta: Lepidoptera). In, Fauna Japonica. Tokyo: Keigaku Publ. 327 pp., 95 pl.
1978. Amphitheridae (Lepidoptera): four new species from Asia, *Telethera blepharacma* Meyrick new to Japan and Formosa, and *Sphenograptis* Meyrick transferred to the family. Bull. Univ. Osaka Pref., (B) 30:1-17.

Newton, P. J., and C. Wilkinson
1982. A taxonomic revision of the North American species of *Stigmella* (Lepidoptera: Nepticulidae). Syst. Ent. (London), 7:367-463.

Nielsen, E. S.
1980. A cladistic analysis of the Holarctic genera of adelid moths (Lepidoptera: Incurvaroidea [sic]). Ent. Scand., 11:161-178.

Nielsen, E. S., and D. R. Davis
1981. A revision of the Neotropical Incurvariidae s. str., with the description of two new genera and two new species (Lepidoptera: Incurvarioidea). Steenstrupia (Copenhagen), 7:25-57.

Opler, P. A., and D. R. Davis
1981. The leafmining moths of the genus *Cameraria* associated with Fagaceae in California (Lepidoptera: Gracillariidae). Smithson. Contr. Zool., 333:1-58.

Passerin d'Entrèves, P.
1977. Revisione degli scitrididi (Lepidoptera, Scythrididae) palearctici. Boll. Istit. Mus. Zool. Univ. Torino (Turin), 5:57-76.

Pastrana, J. A.
1961. La familia Adelidae (Lep.) en la República Argentina. Physis (Buenos Aires), 12:191-201, 3 pl.

Petersen, G.
1957-58. Die Genitalien der paläarktischen Tineiden (Lepidoptera: Tineidae). Beitr. Ent. (Berlin), 7: 55-176, 338-379, 557-595 (1957); 8:111-118, 398-430 (1958).

Povolný, D.
1967a. Genitalia of some Nearctic and Neotropic members of the tribe Gnorimoschemini (Lep., Gel.). Acta Ent. Mus. Natl. Prague, 37:51-127.
1967b. Die stammgeschichtlichen Beziehungen der Tribus Gnorimoschemini im Weltrahmen (Lep., Gel.). Acta Ent. Mus. Natl. Prague, 37:161-232.

Powell, J. A.
1973. A systematic monograph of New World ethmiid moths (Lepidoptera: Gelechioidea). Smithson. Contr. Zool., 120:1-302.
1980. Evolution of larval food preferences in Microlepidoptera. Ann. Rev. Ent., 25:133-159.

Razowski, J.
1974. Phylogeny and classification of Lepidoptera. Acta. Zool. Cracoviensis (Krakow), 19:1-18.

Razowski, J., and J. Wojtusiak
1978. Family-group taxa of Adeloidea (Lepidoptera). Polski Pismo Ent., 48:3-18.

Riedl, T.
1969. Matériaux pour la connaissance des Momphidae paléarctiques (Lepidoptera). Part IX. Revue des Momphidae européenes et compris quelques espèces d'Afrique du Nord et du Proche-Orient. Polski Pismo Ent., 39:635-919. [Momphidae sensu Riedl = Cosmopterigidae, Agonoxenidae and Momphidae].

Robinson, G. S.
1976. A taxonomic revision of the Tinissinae of the World (Lepidoptera: Tineidae). Bull. Br. Mus. (Nat. Hist.), Ent., 32:253-300.
1977. A taxonomic revision of the genus *Callipielus* Butler (Lepidoptera: Hepialidae). Bull. Br. Mus. (Nat. Hist.), Ent., 35:103-121, 6 pl.
1979. Clothes-moths of the *Tinea pellionella* complex: a revision of the world's species (Lepidoptera: Tineidae). Bull. Br. Mus. (Nat. Hist.), Ent., 38:57-128.

Robinson, G. S., and E. S. Nielsen
1983. Ghost-moths of southern South America. Entomonog. (Klampenborg), 4:1-192.

Sattler, K.
1960. Generische Gruppierung der europäischen Arten der Sammelgattung *Gelechia* (Lepidoptera, Gelechiidae) auf Grund der Untersuchungen der männlichen und weiblichen Genitalarmaturen. Dtsch. Ent. Zeit. (Berlin), 7:10-118.
1973. A catalogue of the family-group and genus-group names of the Gelechiidae, Holcopogonidae, Lecithoceridae, and Symmocidae (Lepidoptera). Bull. Br. Mus. (Nat. Hist.), Ent., 28:153-282.

Sattler, K., and W. G. Tremewan
1974. A catalogue of the family-group and genus-group names of the Coleophoridae (Lepidoptera). Bull. Br. Mus. (Nat. Hist.), Ent., 30:184-214.
1978. A supplementary catalogue of the family-group and genus-group names of the Coleophoridae (Lepidoptera). Bull. Br. Mus. (Nat. Hist.), Ent., 37:73-96.

Scoble, M. J.
1982. A pectinifer in the Nepticulidae (Lepidoptera) and its phylogenetic implications. Ann. Transvaal Mus., 33:123-129.
1983. A revised cladistic classification of the Nepticulidae (Lepidoptera) with descriptions of new taxa mainly from South Africa. Transvaal Mus. Monog., 2:1-105.

Toll, S.
1956. Glyphipterygidae. In, Klucze do oznaczania owadów Polski, 27 (Lepidoptera), 39:1-36 Warsaw. [In Polish].
1962. Materialien zur Kenntnis der paläarktischen Arten der Familie Coleophoridae (Lepidoptera). Polski Pismo Ent., 7:577-720, pl. 1-133.

Traugott-Olsen, E., and E. S. Nielsen
1977. The Elachistidae (Lepidoptera) of Fennoscandia and Denmark. In, Fauna Ent. Scand., 6:1-299.

Vári, L.
1961. Lithocolletidae. In, South African Lepidoptera. Volume 1. Transvaal Mus. Mem., 12:1-238, 23 pl.

Vázquez Garcia, L.
1941-[44]. Estudio monográfico de los Psychidae de Mexico. Anal. Inst. Biol. Univ. Mexico, 12:295-310, 3 pl. (1941); 13:257-300, 4 pl. (1942 [1944]).

Wilkinson, C., and P. J. Newton
1981. The microlepidopteran genus *Ectoedemia* Busck (Nepticulidae) in North America. Tijds. Ent. (Amsterdam), 124-27-92, 93-110.

Wojtusiak, J.
1978. Heliozelidae. In, Klucze do oznaczania owadów Polski, 27 (Lepidoptera), 94:1-18. Warsaw. [In Polish].

Zagulajev, A. K.
1960-79. Tineidae. In, Fauna USSR, Lepidoptera, 4(3):1-267 (1960); 4(2):1-424 (1964); 4(4):1-126 (1973); 4(6):1-408 (1979). Leningrad: Akad. Nauk USSR. [In Russian].
1973. On phylogeny in the superfamily Tineoidea (Lepidoptera). Trudy Vsesoy. Ent. Obshch. (Moscow), 56:170-183. [In Russian].

Zimmerman, E. C.
1958-78. Lepidoptera. In, Insects of Hawaii, 7:1-542 (1958); 8:1-456 (1958); 9:1-1876 (1978). Honolulu: Univ. Hawaii Press.

Checklist of Neotropical Lepidoptera

Part 1

GENERIC SYNOPSIS

LEPIDOPTERA

Suborder Zeugloptera

Micropterigoidea

1. MICROPTERIGIDAE

Hypomartyria Kristensen & Nielsen, 1982
Squamicornia Kristensen & Nielsen, 1982

Suborder Heterobathmiina

Heterobathmioidea

2. HETEROBATHMIIDAE

Heterobathmia Kristensen & Nielsen, 1979

Suborder Glossata

Infraorder Neopseustina

Neopseustoidea

3. NEOPSEUSTIDAE

Synempora Davis & Nielsen, 1980
Apoplania Davis, 1975

Infraorder Exoporia

Hepialoidea

4. NEOTHEORIDAE

Neotheora Kristensen, 1978

5. HEPIALIDAE

Dalaca Walker, 1856
Dolaca Druce, 1901, missp.
Huapina Bryk, 1945
Maculella Viette, 1950
Toenga Tindale, 1954
Callipielus Butler, 1882
Stachyocera Ureta, 1957
Blanchardina Viette, 1950
Calada Nielsen & Robinson, 1983
Puermytrans Viette, 1951
Parapielus Viette, 1949
Lossbergiana Viette, 1950
Andeabatis Nielsen & Robinson, 1983
Druceiella Viette, 1949
Trichophassus Le Cerf, 1919
Phassus Walker, 1856
Schausiana Viette, 1949
Aplatissa Viette, 1953
Pfitzneriana Viette, 1952
Pseudophassus Pfitzner, 1938, preocc.
Aepytus Herrich-Schäffer, [1858]
Subgenus **Hampsoniella** Viette, 1949
Subgenus **Pseudodalaca** Viette, 1949
Subgenus **Gymelloxes** Viette, 1952
Subgenus **Alloaepytus** Viette, 1952
Subgenus **Aepytus** Herrich-Schäffer, [1858]
Subgenus **Thiastyx** Viette, 1951
Subgenus **Schaefferiana** Viette, 1950
Subgenus **Paragorgopis** Viette, 1951
Subgenus **Hepialyxodes** Viette, 1951
Subgenus **Xytrops** Viette, 1950
Subgenus **Cibyra** Walker, 1856
Subgenus **Lamelliformia** Viette, 1951
Subgenus **Tricladia** R. Felder, 1874
Pseudophassus R. Felder, 1874
Parana Viette, 1949
Subgenus **Pseudophilaenia** Viette, 1950
Subgenus **Philaenia** Kirby, 1892
Subgenus **Yleuxas** Viette, 1951
Phialuse Viette, 1961
Roseala Viette, 1950
"Dalaca" auct. (not Walker, 1856)
Pfitzneriella Viette, 1950

6. PALAEOSETIDAE

Osrhoes Druce, 1900

Infraorder Heteroneura

Division Monotrysia

Nepticuloidea

7. NEPTICULIDAE

Nepticulinae

Tribe Nepticulini

Stigmella Schrank, 1802
Nepticula Heyden, 1843
Stigmalla Herrich-Schäffer, 1853, missp.
Dysnepticula Börner, 1925
Johanssonia Borkowski, 1972

Tribe Trifurculini

Enteucha Meyrick, 1915

8. OPOSTEGIDAE

Opostega Zeller, 1839

9. TISCHERIIDAE

Tischeria Zeller, 1839
Evexia Gistel, 1847
Philodoxa Gistel, 1848
Coptotriche Walsingham, 1890
Tisheria Busck [1903], missp.

Incurvarioidea

10. INCURVARIIDAE

Basileura Nielsen & Davis, 1981
Simacauda Nielsen & Davis, 1981

11. CECIDOSIDAE

Cecidoses Curtis, 1835
Clistoses Kieffer & Jörgensen, 1910
Eucecidoses Brèthes, 1916
Oliera Brèthes, 1916
Ridiaschinia Brèthes, 1916
Dicranoses Kieffer & Jörgensen, 1910

12. ADELIDAE

Ceromitia Zeller, 1854
Trichorrhabda Meyrick, 1912, **n. syn.**
Adela Latreille, 1796
Capillaria Haworth, 1828
Metallitis Sodoffsky, 1837
Aedilis Gistel, 1873
Dicte Chambers, 1873

13. PRODOXIDAE

Tegeticula Zeller, 1873
Pronuba Riley, 1872, preocc.
Promiba Kirbey, 1874, missp.
Thia H. Edwards, 1888, preocc.
Thelethia Dyar, 1893
Valentina Coolidge, 1909, preocc.
Parategeticula Davis, 1967
Prodoxus Riley, 1880
Prodoxoides Nielsen & Davis, 1984

14. HELIOZELIDAE

Antispila Hübner, [1825]
Coptodisca Walsingham, 1895
Aspidisca Clemens, 1860, preocc.
Heliozela Herrich-Schäffer, 1853
Dyselachista Spuler, 1910
Lamprozela Meyrick, 1916
Monachozela Meyrick, 1931
Phanerozela Meyrick, 1921

Division Ditrysia

Section Tineina

Tineoidea

15. TINEIDAE

Acrolophus Poey, 1832
Anaphora Clemens, 1859
Daulia Walker, 1863, preocc.
Derchis Walker, 1863
Eddara Walker, 1863, preocc.
Hibita Walker, 1863
Tirasia Walker, 1863
Bazira Walker, 1864
Phlongia Walker, 1864
Urbara Walker, 1864
Tachasara Walker, 1865
Eutheca Grote, 1881, preocc.
Eulepiste Walsingham, 1882
Pseudoconchylis Walsingham, 1884
Homonymus Walsingham, 1887
Neolophus Walsingham, 1887
Hypoclopus Walsingham, 1887
Thysanoscelis Walsingham, 1887
Thysanoskelis Walsingham, 1887
Ankistrophorus Walsingham, 1887
Caenogenes Walsingham, 1887
Felderia Walsingham, 1887
Ortholophus Walsingham, 1887
Pseudanophora Walsingham, 1887
Sapinella Kirby, 1892
Atopocera Walsingham, 1897
Pilanophora Walsingham, 1897
Thysanosedes Druce, 1901, missp.
Hypocolypus Dyar, [1903], missp.
Anophora Kearfott, 1903, missp.
Brachysymbola Meyrick, 1912
Orothyntis Meyrick, 1913
Apoclisis Walsingham, 1914
Psephocrita Meyrick, 1919
Xerocausta Meyrick, 1928, **n. syn.**
Anaphorina Strand, 1932
Amydria Clemens, 1859
Casape Walker, 1864
Amadria Chambers, 1875, missp.
Amadrya Chambers, 1875, missp.
Amydrya Kearfott, 1903, missp.
Neomeristis Meyrick, 1919, **n. syn.**
Antipolistes Forbes, 1933
Atticonviva Busck, 1934
Augolychna Meyrick, 1922
Axiagasta Meyrick, 1930
Barymochtha Meyrick, 1922
Basanasca Meyrick, 1922
Brithyceros Meyrick, 1932
Bythocrates Meyrick, 1919
Choropleca Durrant, 1914
Cyane Chambers, 1873, preocc.
Diachalastis Meyrick, 1920
Clepticodes Meyrick, 1927
Clinograptis Meyrick, 1932
Cnismorectis Meyrick, 1936
Colpocrita Meyrick, 1930
Comodica Meyrick, 1880
Compsocrita Meyrick, 1922
Cranaodes Meyrick, 1919
Crepidochares Meyrick, 1922
Cryphiotechna Meyrick, 1932
Dasmophora Meyrick, 1919
Demobrotis Meyrick, 1892
Diataga Walsingham, 1914

Dorata Busck, 1904
Pterolonche Walsingham, 1889, preocc.
Dorota Kearfott, 1907, missp.
Drastea Walsingham, 1914
Dyotopasta Busck, 1907
Pseudoxylesthia Walsingham, 1907
Dystopasta McDunnough, 1939, emend.
Dysoptus Walsingham, 1914
Ectinocampa Silvestri 1944
Ephedroxena Meyrick, 1919
Episyrta Meyrick, 1929
Erechthias Meyrick, 1880
Ereunetis Meyrick, 1880
Decadarchis Meyrick, 1886
Nesoxena Meyrick, 1929
Amphisyncentris Meyrick, 1933
Caryolestis Meyrick, 1934
Triadogona Meyrick, 1937
Anemerarcha Meyrick, 1937
Erechtias Ghesquière, 1940
Tinexotaxa Gozmány, 1968
Neodecadarchis Zimmerman, 1978
Lepidobregma Zimmerman, 1978
Pantheus Zimmerman, 1978
Exoncotis Meyrick, 1919
Homodoxus Walsingham, 1914
Homosetia Clemens, 1863
Homostinea Dietz, 1905
Homotinea Meyrick, 1932, emend.
Hormantris Meyrick, 1927
Hybroma Clemens, 1862
Infurcitinea Spuler, 1910
Atinea Amsel, 1954
Microtinea Amsel, 1954
Isocorypha Dietz, 1905
Socorypha Busck, 1909, missp.
Leptochersa Meyrick, 1919
Lepyrotica Meyrick, 1921
Leucophasma Walsingham, 1897
Lindera Blanchard, 1852
Palpula Blanchard, 1852, preocc.
Safra Walker, 1864, preocc.
Chresotes Butler, 1881
Paraneura Dietz, 1905
Cervitinea Amsel, 1956
Lipomerinx Walsingham, 1914
Lithopsaestis Meyrick, 1932
Mea Busck, 1906
Progona Dietz, 1905, preocc.
Monopis Hübner, [1825]
Blabophanes Zeller, 1852
Hyalospila Herrich-Schäffer, 1853
Rhitia Walker, 1864
Eusynopa Lower, 1903
Myrmecozela Zeller, 1852
Psecadioides Butler, 1881
Promasia Chrètien, 1905
Protolophe Rebel, 1915
Mythoplastis Meyrick, 1919
Nemapogon Schrank, 1802
Diaphthirusa Hübner, [1825]
Neomeristis Meyrick, 1919
Niditinea Petersen, 1957
Tineidia Zagulajev, 1960
Oenoe Chambers, 1874
Oinophila Stephens, 1848
Oenophila Zeller, 1853, missp.

Opogona Zeller, 1853
Lozostoma Stainton, 1859
Conchyliospila Wallengren, 1861
Cachura Walker, 1864
Dendroneura Walsingham, 1892
Hieroxestis Meyrick, 1892
Exala Meyrick, 1912
Opsodoca Meyrick, 1919
Otochares Meyrick, 1919
Pachydyta Meyrick, 1922
Pachdyta Meyrick, 1922, missp.
Palaephatus Butler, 1883
Panthytarcha Meyrick, 1922
Pedaliotis Meyrick, 1930
Perilicmetis Meyrick, 1932
Phaeoses Forbes, 1922
Phereoeca Hinton & Bradley, 1956
Polypsecta Meyrick, 1930
Pompostolella Fletcher, 1940
Pompostola Meyrick, 1927, preocc.
Praeacedes Amsel, 1954
Titaenoses Hinton & Bradley, 1956
Proboloptila Meyrick, 1921
Protodarcia Forbes, 1931
Ptilopsaltis Meyrick, 1935
Scardia Treitschke, 1830
Agarica Sodoffsky, 1837
Morophaga Herrich-Schäffer, 1853
Atabyria Snellen, 1884
Osphretica Meyrick, 1910
Microscardia Amsel, 1952
Setiarcha Meyrick, 1932
Setomorpha Zeller, 1852
Epilegis Dietz, 1905
Apotomia Dietz, 1905
Semiota Dietz, 1905
Trisyntopa Lower, 1918
Syncraternis Meyrick, 1922
Syrmologa Meyrick, 1919
Taeniodictys Forbes, 1933
Tetrapalpus Davis, 1972
Tinea Linnaeus, 1758
Autoses Hübner, [1825]
Acedes Hübner, [1825]
Edosa Walker, 1866
Chrysoryctis Meyrick, 1886
Dystinea Börner, 1925
Tiquadra Walker, 1863
Oscella Walker, 1864
Manchana Walker, 1866
Ventia Walker, 1866
Acureuta Zeller, 1877
Trichophaga Ragonot, 1894
Trierostola Meyrick, 1932
Xylesthia Clemens, 1859
Xylestia Dyar, [1903], missp.
Xystrologa Meyrick, 1919
Achanodes Meyrick, 1922
Syrrhoaula Meyrick, 1932
Zonochares Meyrick, 1922
Zymologa Meyrick, 1919

16. PSYCHIDAE

Penestoglossinae

Amiantastis Meyrick, 1932
Autocnaptis Meyrick, 1935
Plumana Busck, 1911
Homilostola Meyrick, 1917
Perisceptis Meyrick, 1931
Pterogyne Davis, 1975

Psychinae

Epichnopterix Hübner, [1825]
Epichnopteryx Heylaerts, 1875, missp.
Paucivena Davis, 1975
Prochalia Barnes & McDunnough, 1913
Metaxypsyche Davis, 1975
Naevipenna Davis, 1964
Cryptothelea Duncan, 1841
Platoeceticus Packard, 1869
Cryptotheles Costa Lima, 1945, missp.
Dendropsyche F. M. Jones, 1926
Lumacra Davis, 1964
Curtorama Davis, 1964
Astala Davis, 1964

Oiketicinae

Thantopsyche Butler, 1882
Animula Herrich-Schäffer, 1856
Subgenus **Animula** Herrich-Schäffer, 1856
Subgenus **Antipenna** Davis, 1964
Biopsyche Dyar, 1905
Oiketicus Guilding, 1827
Oeketicus Lefebre, 1842, missp.
Oeceticus Herrich-Schäffer, 1858, missp.
Oiketikus Seitz, 1919, missp.
Oicocestis Köhler, 1924, missp.
Oiceticus Lahille, 1926, missp.
Subgenus **Paraoiketicus** Davis, 1964
Subgenus **Oiketicus** Guilding, 1827
Thyridopteryx Stephens, 1834
Hymenopsyche Grote, 1865

17. ARRHENOPHANIDAE

Arrhenophanes Walsingham, 1913
Cnissostages Zeller, 1863
Ecpathophanes Bradley, 1951
Harmaclona Busck, 1914
Ptychoxena Meyrick, 1916

18. AMPHITHERIDAE

Dasycarea Zeller, 1877

19. LYONETIIDAE

Bedelliinae

Euprora Busck, 1906
Philonome Chambers, 1874
Phillonome Chambers, 1875, missp.
Eurynome Chambers, 1875, preocc.
Busckia Dyar, [1903]
Bedellia Stainton, 1849

Lyonetiinae

Lyonetia Hübner, [1825]
Argyromiges Curtis, 1829
Argyromis Stephens, 1829
Lyoneta Matsumura, 1931, missp.

Cemiostominae

Leucoptera Hübner, [1825]
Cemiostoma Zeller, 1848
Perileucoptera Silvestri, 1943
Compsoschema Walsingham, 1897
Hapalothyma Meyrick, 1920
Thomictis Meyrick, 1920

Bucculatriginae

Bucculatrix Zeller, 1848
Ceroclastis Zeller, 1848
Bacculatrix Flint, Noble, & Shaw, 1978, missp.

20. GRACILLARIIDAE

Gracillariinae

Caloptilia Hübner, [1825]
Poeciloptilia Hübner, [1825]
Ornix Treitschke, 1833
Coriscium Zeller, 1839
Antiolopha Meyrick, 1894
Eucosmophora Walsingham, 1897
Neurostrota Ely, 1918
Neurostrata Ely, 1918, missp.
Parectopa Clemens, 1860
Euspilopteryx Zeller, 1847, emend., misid.
Euspilapteryx Spuler, 1910, missp.
Neurobathra Ely, 1918
Penica Walsingham, 1914
Parornix Spuler, 1910
Chilocampyla Busck, 1900
Dialectica Walsingham, 1897
Eutrichocnemis Spuler, 1910
Acrocercops Wallengren, 1881
Cuphodes Meyrick, 1897
Phrixosceles Meyrick, 1908
Spanioptila Walsingham, 1897
Leucanthiza Clemens, 1859
Marmara Clemens, 1863
Aesyle Chambers, 1875

Lithocolletinae

Cremastobombycia Braun, 1908
Phyllonorycter Hübner, [1825]
Lithocolletis Hübner, [1825]
Eucestis Hübner, [1825]
Lithcolletes Matsumura, 1907, missp.
Phyllorycter Walsingham, 1914, missp.
Cameraria Chapman, 1902
Porphyrosela Braun, 1908

Phyllocnistinae

Phyllocnistis Zeller, 1848
Phylloenistis Chambers, 1875, missp.
Phylloetis Chambers, 1876, missp.
Phyllocnitis Busck, 1900, missp.

Gelechioidea

21. OECOPHORIDAE

Depressariinae

Tribe Depressariini

Exaeretia Stainton, 1849
Depressariodes Turati, 1924
Martyrhilda Clarke, 1941
Talitha Clarke, 1978

Tribe Amphisbatini

Afdera Clarke, 1978
Ancipita Busck, 1914
Anthrinacia Walsingham, 1911
Auxotricha Meyrick, 1931
Chariphylla Meyrick, 1921
Comotechna Meyrick, 1920
Compsistis Meyrick, 1888
Coptotelia Zeller, 1863
Hyphypena Warren, 1889
Costoma Busck, 1914
Phalarotarsa Meyrick, 1924
Cryptolechia Zeller, 1852
Doina Clarke, 1978
Doshia Clarke, 1978
Ectaga Walsingham, 1912
Erithyma Meyrick, 1914
Eupragia Walsingham, 1911
Filinota Busck, 1911
Lupercalia Busck, 1912
Mnesichara Walsingham, 1912
Gnathotona Meyrick, 1931
Gonada Busck, 1911
Gonionota Zeller, 1877
Habrophylax Meyrick, 1931
Hamadera Busck, 1914
Hastamea Fletcher, 1929, repl. name
Hasta Busck, 1911, preocc.
Himotica Meyrick, 1912
Hypercallia Stephens, 1834
Agriocoma Zeller, 1877
Brachyplatea Zeller, 1877
Eumimographe Dognin, 1905
Idiocrates Meyrick, 1909
Lucyna Clarke, 1978
Machimia Clemens, 1860
Maesara Clarke, 1968
Melaneulia Butler, 1883
Muna Clarke, 1978
Nedenia Clarke, 1978
Nematochares Meyrick, 1931
Osmarina Clarke, 1978
Palinorsa Meyrick, 1924
Perzelia Clarke, 1978
Philtronoma Meyrick, 1914
Pholcobates Meyrick, 1931
Phytomimia Walsingham, 1912
Pisinidea Butler, 1883
Profilinota Clarke, 1973
Pseudocentris Meyrick, 1921
Psilocorsis Clemens, 1860
Paepia Walker, 1864
Hagno Chambers, 1872
Psittacastis Meyrick, 1909
Necedes Walsingham, 1912
Rhindoma Busck, 1914
Scoliographa Meyrick, 1916
Taruda Walker, 1864
Ecliptoloma Zeller, 1877
Trycherodes Meyrick, 1914
Teratomorpha Walsingham, 1912, preocc.
Zemiocrita Meyrick, 1933

Ethmiinae

Erysiptila Meyrick, 1914
Cyrictodes Meyrick, 1926
Ethmia Hübner, [1819]
Psecadia Hübner, [1825]
Anesychia Hübner, [1825]
Disthymnia Hübner, [1825]
Melanoleuca Stephens, 1829
Aedia Duponchel, 1836
Chalybe Duponchel, 1836
Azinis Walker, 1863
Tamarrha Walker, 1864
Ceratophysetis Meyrick, 1887
Theoxenia Walsingham, 1887
Babaiaxa Busck, 1902
Wiltshireia Amsel, 1949
Macrocirca Meyrick, 1931

Peleopodinae

Durrantia Busck, 1908
Dolidiria Busck, 1912
Peleopoda Zeller, 1877
Theatria Walsingham, 1912
Pseuderotis Clarke, 1956
Schistonoea Forbes, 1931

Stenomatinae

Anadasmus Walsingham, 1897
Anapatris Meyrick, 1932
Antaeotricha Zeller, 1854
Mesoptycha Zeller, 1854
Brachiloma Clemens, 1863
Harpalyce Chambers, 1874
Ide Chambers, 1880
Antoeotricha Walsingham, 1881, missp.
Aedemoses Walsingham, 1912
Athleta Walsingham, 1912
Prasolithites Meyrick, 1912
Aphanoxena Meyrick, 1915
Psephomeres Meyrick, 1916
Eumiturga Meyrick, 1925
Baeonoma Meyrick, 1916
Catarata Walsingham, 1912

Cerconota Meyrick, 1915
Pomphocrita Meyrick, 1930
Cerconata Busck, 1935, missp.
Chlamydastis Meyrick, 1916
Ptilogenes Meyrick, 1917
Diastoma Möschler, 1882
Energia Walsingham, 1912
Falculina Zeller, 1877
Gonioterma Walsingham, 1897
Hyalopseustis Meyrick, 1925
Lethata Duckworth, 1964
Loxotoma Zeller, 1854
Menesta Clemens, 1860
Mothonica Walsingham, 1912
Mysaromima Meyrick, 1926
Orphnolechia Meyrick, 1909
Parascaeas Meyrick, 1936
Paraspastis Meyrick, 1915
Petalothyrsa Meyrick, 1931
Petasanthes Meyrick, 1925
Phelotropa Meyrick, 1915
Promenesta Busck, 1914
Rectiostoma Becker, 1982, repl. name
Setiostoma Zeller, 1875, preocc.
Rhodonassa Meyrick, 1915
Stenoma Zeller, 1839
Auxocrossa Zeller, 1854
Auxocrassa Walsingham, 1882, missp.
Thioscelis Meyrick, 1909
Timocratica Meyrick, 1912
Lychnocrates Meyrick, 1926
Zetesima Walsingham, 1912

Oecophorinae

Tribe Oecophorini

Acartophila Meyrick, 1932
Aliciana Clarke, 1978
Altiura Clarke, 1978
Alynda Clarke, 1978
Aniuta Clarke, 1978
Arctopoda Butler, 1883
Polypseustis Dognin, 1908
Atelosticha Meyrick, 1883
Crypsynarthra Lower, 1901
Atha Clarke, 1978
Atopotorna Meyrick, 1932
Borkhausenia Hübner, [1825]
Amaurosetia Stephens, [1835]
Pseudatemelia Rebel, 1910
Brymblia Hodges, 1974
Callistenoma Butler, 1883
Cecidolechia Kieffer & Jörgensen, 1910
Cecidolechia Strand, 1911, redesc.
Chezala Walker, 1864
Corita Clarke, 1978
Decantha Busck, 1914
Deia Clarke, 1978
Despina Clarke, 1978
Dinotropa Meyrick, 1916
Dita Clarke, 1978
Doliotechna Meyrick, 1914
Dysgnorima Zeller, 1877
Endrosis Hübner, [1825]
Eomichla Meyrick, 1916
Astoxena Meyrick, 1930
Eraina Clarke, 1978
Exosphrantis Meyrick, 1931
Halimarmara Meyrick, 1931
Harpella Schrank, 1802
Heliostibes Zeller, 1874
Hofmannophila Spuler, 1910
Hyperskeles Butler, 1883
Ifeda Hodges, 1966
Inga Busck, 1908
Doxa Walsingham, 1912
Lysigrapha Meyrick, 1914
Pelomimas Meyrick, 1924
Orsimacha Meyrick, 1914
Siderograptis Meyrick, 1920
Phanerodoxa Meyrick, 1921
Epimoryctis Meyrick, 1930
Horomeristis Meyrick, 1931
Agriotorna Meyrick, 1931
Irenia Clarke, 1978
Lelita Clarke, 1978
Lygronoma Meyrick, 1913
Macarocosma Meyrick, 1931
Mathildana Clarke, 1941
Melochrysis Meyrick, 1916
Odonna Clarke, 1982
Pachyphoenix Butler, 1883
Tyriomorpha Meyrick, 1918
Mattea Duckworth, 1966
Philomusaea Meyrick, 1931
Philomusea Clarke, 1978, missp.
Pseudoecophora Staudinger, 1899
Pycnotarsa Meyrick, 1920
Pyramidobela Braun, 1923
Idioptila Meyrick, 1927
Retha Clarke, 1978
Revonda Clarke, 1978
Stasixena Meyrick, 1930
Struthoscelis Meyrick, 1913
Teresita Clarke, 1978
Thaumatolita Walsingham, 1912
Theama E. M. Hering, 1958
Utilia Clarke, 1978
Zymrina Clarke, 1978

Tribe Stathmopodini

Arauzona Walker, [1865]
Snellenia Walsingham, 1889
Stathmopoda Herrich-Schäffer, 1853
Boocara Butler, 1880
Placostola Meyrick, 1887
Erineda Busck, 1909
Agrioscelis Meyrick, 1913
Kakivoria Nagano, 1916
Tinaegeria Walker, 1856

Tribe Pleurotini

Pleurota Hübner, [1825]
Eupleuris Hübner, [1825]
Holoscolia Zeller, 1839
Protasis Herrich-Schäffer, 1853

22. ELACHISTIDAE

Elachistinae

Dicranoctetes Braun, 1918
Donacivola Busck, [1934]
Elachista Treitschke, 1833
Aphelosetia Stephens, 1834
Cycnodia Herrich-Schäffer, 1853
Phigalia Chambers, 1875, preocc.
Atachia Wocke, 1876
Neaera Chambers, 1880, preocc.
Hecista Wallengren, 1881
Aphigalia Dyar, 1903
Irenicodes Meyrick, 1919
Euproteodes Viette, 1954

23. BLASTOBASIDAE

Symmocinae

Eupolella Fletcher, 1940, repl. name
Eupolis Meyrick, 1923, preocc.
Glyphidocera Walsingham, 1892
Harpagandra Meyrick, 1918
Oecia Walsingham, 1897
Macroceras Staudinger, 1876, preocc.
Sceptea Walsingham, 1911
Sceptia McDunnough, 1939, missp.
Symmoca Hübner, [1825]
Simoca Weiler, 1877, missp.
Parasymmoca Rebel, 1903
Asarista Meyrick, 1935
Conquassata Gozmány, 1957
Symmoletria Gozmány, 1963

Blastobasinae

Tribe Blastobasini

Auximobasis Walsingham, 1892
Blastobasis Zeller, 1855
Epistetus Walsingham, 1894
Protesis Walsingham, 1908
Holcocera Clemens, 1863
Catacrypsis Walsingham, 1907
Cynotes Walsingham, 1907
Prosodica Walsingham, 1907
Iconisma Walsingham, 1897
Metallocrates Meyrick, 1930
Valentinia Walsingham, 1907

Tribe Pigritiini

Pigritia Clemens, 1860

24. COLEOPHORIDAE

Coleophorinae

Coleophora Hübner, [1822]
Eupista Hübner, [1825]
Apista Hübner, [1825]
Haploptilia Hübner, [1825]
Porrectaria Haworth, 1828
Damophila Curtis, 1832
Astyages Stephens, 1834
Metallosetia Stephens, 1834
Casas Wallengren, 1881
Casigneta Wallengren, 1881
Corythangela Meyrick, 1897
Casignetella Strand, 1928
Calaritania Mariani, 1943
Heringiella Börner, 1944
Tolleophora Căpuse, 1971
Ionescumia Căpuse, 1971
Stollia Căpuse, 1971
Razowskia Căpuse, 1971
Orghidania Căpuse, 1971
Frederickoenigia Căpuse, 1971
Suireia Căpuse, 1971
Zagulajevia Căpuse, 1971
Amseliphora Căpuse, 1971
Nemesia Căpuse, 1971
Zangheriphora Căpuse, 1971
Bourgogneja Căpuse, 1971
Aureliania Căpuse, 1971
Bacescuia Căpuse, 1971
Klinzigedia Căpuse, 1971
Vladdelia Căpuse, 1971
Klimeschja Căpuse, 1971
Glaseria Căpuse, 1971
Valvulongia Căpuse, 1971
Falkovitshia Căpuse, 1972
Helopharea Falkovitsh, 1972
Cricotechna Falkovitsh, 1972
Plegmidia Falkovitsh, 1972
Agapalsa Falkovitsh, 1972
Phylloschema Falkovitsh, 1972
Bima Falkovitsh, 1972
Systrophoeca Falkovitsh, 1972
Aporiptura Falkovitsh, 1972
Symphypoda Falkovitsh, 1972
Oedicaula Falkovitsh, 1972
Argyractinia Falkovitsh, 1972
Chnoocera Falkovitsh, 1972
Orthographis Falkovitsh, 1972
Phagolamia Falkovitsh, 1972
Monotemachia Falkovitsh, 1972
Corethropoea Falkovitsh, 1972
Characia Falkovitsh, 1972
Perygra Falkovitsh, 1972
Perygridia Falkovitsh, 1972
Luzulina Falkovitsh, 1972
Carpochena Falkovitsh, 1972
Abaraschia Căpuse, 1973
Amselghia Căpuse, 1973
Ardania Căpuse, 1973
Ascleriductia Căpuse, 1973
Baraschia Căpuse, 1973
Benanderpia Căpuse, 1973
Calcomarginia Căpuse, 1973
Caleophora Căpuse, 1973, missp.
Corothropoea Căpuse, 1973, missp.
Cornulivalvulia Căpuse, 1973
Dumitrescumia Căpuse, 1973
Ecebalia Căpuse, 1973
Globulia Căpuse, 1973
Hamuliella Căpuse, 1973
Helvalbia Căpuse, 1973

Ionnemesia Căpuse, 1973
Kasyfia Căpuse, 1973
Kuznetzovvlia Căpuse, 1973
Latisacculia Căpuse, 1973
Longibacillia Căpuse, 1973
Lucidaesia Căpuse, 1973
Lvaria Căpuse, 1973
Membrania Căpuse, 1973
Metapista Căpuse, 1973
Multicoloria Căpuse, 1973
Neugenvia Căpuse, 1973
Nosyrislia Căpuse, 1973
Ortohgraphis Căpuse, 1973, missp.
Oudejansia Căpuse, 1973
Paravalvulia Căpuse, 1973
Patzakia Căpuse, 1973
Postvinculia Căpuse, 1973
Proglaseria Căpuse, 1973
Quadratia Căpuse, 1973
Rhamnia Căpuse, 1973
Sacculia Căpuse, 1973
Scleriductia Căpuse, 1973
Tollsia Căpuse, 1973
Tuberculia Căpuse, 1973
Ulna Căpuse, 1973
Ductispira Căpuse, 1973
Klimeschjosefia Căpuse, 1975
Tocasta Busck, 1912

Batrachedrinae

Amblytenes Meyrick, 1930
Batrachedra Herrich-Schäffer, 1853
Eustaintonia Spuler, 1910
Chedra Hodges, 1966

25. MOMPHIDAE

Anchimompha Clarke, 1965
Mompha Hübner, [1825]
Laverna Curtis, 1839
Lophoptilus Sircom, 1848
Cyphophora Herrich-Schäffer, 1853
Psacaphora Herrich-Schäffer, 1853
Anybia Stainton, 1854
Wilsonia Clemens, 1864
Leucophryne Chambers, 1875
Synallagma Busck, 1907

26. AGONOXENIDAE

Blastodacninae

Homoeoprepes Walsingham, 1909
Homeoprepes Hodges, 1978, missp.
Microcolona Meyrick, 1897
Nanodacna Clarke, 1964
Nicanthes Meyrick, 1928
Pammeces Zeller, 1863
Panclintis Meyrick, 1929
Prochola Meyrick, 1915
Syntetrernis Meyrick, 1922
Zaratha Walker, 1864

27. COSMOPTERIGIDAE

Antequerinae

Antequera Clarke, 1941
Euclemensia Grote, 1878, repl. name
Hamadryas Clemens, 1864, preocc.

Cosmopteriginae

Anoncia Clarke, 1941
Aphanosara Forbes, 1931
Cosmopterix Hübner, [1825]
Cosmopteryx Zeller, 1839, emend.
Dromiaulis Meyrick, 1922
Ecballogonia Walsingham, 1912
Harpograptis Meyrick, 1925
Labdia Walker, 1864
Limnaecia Stainton, 1851
Lymnaecia Kimball, 1965, missp.
Moriloma Busck, 1912
Pyroderces Herrich-Schäffer, 1853
Anatrachyntis Meyrick, 1915
Sathrobrota Hodges, 1962
"**Scaeosopha**" auct., (not Meyrick, 1914)
Sematoptis Meyrick, 1931
Teladoma Busck, 1932
Triclonella Busck, 1901
Anorcota Meyrick, 1920
Pharmacoptis Meyrick, 1932
Urangela Busck, 1912

Chrysopeleiinae

Ascalenia Wocke, 1876
Eritarbes Walsingham, 1909
Ithome Chambers, 1875
Eriphia Chambers, 1875, preocc.
Leptozestis Meyrick, 1924
Obithome Hodges, 1964
Perimede Chambers, 1874
Periploca Braun, 1919
Stilbosis Clemens, 1860
Aeaea Chambers, 1874
Amaurogramma Braun, 1919
Walshia Clemens, 1864

28. SCYTHRIDIDAE

Scythris Hübner, [1825]
Butalis Treitschke, 1833, preocc.
Arotrura Walsingham, 1888
Colinita Busck, 1907

29. GELECHIIDAE

Anomologinae

Isophrictis Meyrick, 1917
Megacraspedus Zeller, 1839
Neda Chambers, 1874, preocc.
Pycnobathra Lower, 1901
Autoneda Busck, 1903, repl. name
Toxoceras Chrétien, 1915
Megacraspedas Barnes & McDunnough, 1917, missp.

Monochroa Heinemann, 1870
Catabrachmia Rebel, 1909
Nealyda Dietz, 1900

Gelechiinae

Aristotelia Hübner, [1825]
Ergatis Heinemann, 1870, preocc.
Isochasta Meyrick, 1886
Eucatoptus Walsingham, 1897
Arla Clarke, 1942
Aroga Busck, 1914
Aruga Janse, 1958, missp.
Arogalea Walsingham, 1910
Barticeja Povolný, 1967
Calliprora Meyrick, 1914
Chionodes Hübner, [1825]
Chionoda Hübner, [1826], missp.
Coleotechnites Chambers, 1880
Evagora Clemens, 1860, preocc.
Eidothea Chambers, 1873, preocc.
Eidothoa Chambers, 1873, missp.
Coleotechnistes Riley, 1891, missp.
Eucordylea Dietz, 1900
Pulicalvaria T. N. Freeman, 1963
Deltophora Janse, 1950
Ephysteris Meyrick, 1908
Microcraspedus Janse, 1958
Ephystereris Janse, 1960, missp.
Echinoglossa Clarke, 1965
Euchionodes Clarke, 1950
Eudactylota Walsingham, 1911
Evippe Chambers, 1873
Phaetusa Chambers, 1875, preocc.
Tholerostola Meyrick, 1917
Exoteleia Wallengren, 1881
Paralechia Busck, 1903
Heringia Spuler, 1910, preocc.
Heringiola Strand, 1917, repl. name
Faculta Busck, 1939
Fascista Busck, 1939
Friseria Busck, 1939
Gelechia Hübner, [1825]
Guenea Bruand, 1850
Galechia Desmarest, [1857], missp.
Gelschia Nowicki, 1865, missp.
Cirrha Chambers, 1872
Oeseis Chambers, 1875
Gelecia Watt, 1920, missp.
Gnorimoschema Busck, 1900
Gnorimochema Dyar, [1903], missp.
Tuta Kieffer & Jörgensen, 1910
Gonorimoschema Deurs, 1954, missp.
Gnorrimoschema Hartig, 1964, missp.
Lerupsia Riedl, 1965
Neoschema Povolný, 1967
Larupsia Soffner, 1967, missp.
Hapalosaris Meyrick, 1917
Keiferia Busck, 1939
Lita Treitschke, 1833
Locharca Meyrick, 1923
Neodactylota Busck, 1903
Orsotricha Meyrick, 1914
Phthorimaea Meyrick, 1902
Phtyorimaea Turner, 1919, missp.
Phthorimoea Povolný & Zakopal, 1951, missp.
Pthorimaea Issiki, 1957, missp.
Phthorimea Diakonoff, [1968], missp.
Phtorimea Oei-Dharma, 1969, missp.
Polyhymno Chambers, 1874
Copocercia Zeller, 1877
Oegoconiodes Matsumura, 1931
Oegoconides Neave, 1940, missp.
Pseudarla Clarke, 1965
Ptycerata Ely, 1910
Scrobipalpopsis Povolný, 1967
Subgenus **Ptycerata** Ely, 1910
Subgenus **Scrobischema** Povolný, 1979
Recurvaria Haworth, 1828
Telea Stephens, 1834, preocc.
Aphanaula Meyrick, 1895
Hinnebergia Spuler, 1910
Microlechia Turati, 1924
Schistophila Chrétien, 1899
Scrobipalpula Povolný, 1964
Subgenus **Eurysacca** Povolný, 1967
Subgenus **Magnifascia** Povolný, 1967
Subgenus **Scrobipalpula** Povolný, 1964
Sophronia Hübner, [1825]
Stegasta Meyrick, 1904
Stomopteryx Heinemann, 1870
Inotica Meyrick, 1913
Instica Sharp, 1915, missp.
Acraeologa Meyrick, 1921
Stomopterix Turati, 1922, missp.
Stromopteryx Pierce & Metcalfe, 1935, missp.
Symmetrischema Povolný, 1967
Taygete Chambers, 1873
Telphusa Chambers, 1872
Adrasteia Chambers, 1872
Adrastia Kirby, 1874, emend.
Geniadophora Walsingham, 1897
Telephusa Beirne, 1938, missp.
Thiotricha Meyrick, 1886
Reuttia O. Hofmann, 1898
Thiotrica Inoue, 1954, missp.
Thiothricha Hartig, 1956, missp.
Tildenia Povolný, 1967

Anacampsinae

Anacampsis Curtis, 1827
Anacompsis Desmarest, [1857], missp.
Tachyptilia Heinemann, 1870
Tuchyptilia Kirby, 1871, missp.
Tachiptilia Chambers, 1878, missp.
Tachyptilix Hartmann, 1880, missp.
Agriastis Meyrick, 1914
Agriastsi Busck, 1919, missp.
Tachoptilia Daltry, 1926, missp.
Trachyphilia Le Marchand, 1947, missp.
Trachyptilia Le Marchand, 1947, missp.
Battaristis Meyrick, 1914
Duvita Busck, 1916
Compsolechia Meyrick, 1918
Holophysis Walsingham, 1910
Hoplophysis McDunnough, 1939, missp.
Strobisia Clemens, 1860
Systasiota Walsingham, 1910
Untomia Busck, 1906

Chelariinae

Crasimorpha Meyrick, 1923
Epilechia Busck, 1939
Haplochela Meyrick, 1923
Hypatima Hübner, [1825]
Chelaria Haworth, 1828
Hypatina Stephens, 1835, missp.
Allocota Meyrick, 1904, preocc.
Cynestomorpha Meyrick, 1904
Deuteroptila Meyrick, 1904
Cymatomorpha Meyrick, 1904
Semodictis Meyrick, 1909
Allocotaniana Strand, 1913, repl. name
Episacta Turner, 1919
Cellaria Neave, 1939, missp.
Cheleria Lhomme, [1948], missp.
Pectinophora Busck, 1917
Pessograptis Meyrick, 1923
Porpodryas Meyrick, 1920
Prostomeus Busck, 1903
Semophylas Meyrick, 1932
Sitotroga Heinemann, 1870
Silotroga Kirby, 1871, missp.
Nesolechia Meyrick, 1921
Syngenomictis Meyrick, 1927
Sitotrogus Matsumura, 1931, missp.
Sitotrega Borg, 1932, missp.
Sititroga Costa Lima, 1945, missp.

Dichomeridinae

Acompsia Hübner, [1825]
Acampsia Westwood, 1840, missp.
Accompsia Bruand, 1850, missp.
Brachycrossata Heinemann, 1870
Brachicrossata Hartmann, 1880, missp.
Anorthosia Clemens, 1860
Sagaritis Chambers, 1872, preocc.
Anorthodisca Gaede, 1937, missp.
Brachmia Hübner, [1825]
Braclunia Stephens, 1834, missp.
Cladodes Heinemann, 1870, preocc.
Ceratophora Heinemann, 1870, preocc.
Eudodacles Snellen, 1889, repl. name
Aulacomima Meyrick, 1904
Apethistis Meyrick, 1908
Brachyacma Meyrick, 1886
Lathontogenus Walsingham, 1897
Paraspistes Meyrick, 1905
Lipatia Busck, 1910
Paraspistis Busck, 1914, missp.
Brachyaema Povolný, 1964, missp.
Lathontogonus Diakonoff, [1968], missp.
Brachiacma Common, 1970, missp.
Lathontogenes Hodges, 1983, missp.
Cotyloscia Meyrick, 1923
Cymotricha Meyrick, 1923
Oxysactis Meyrick, 1923
Dichomeris Hübner, [1818]
Oxybelia Hübner, [1825]
Rhinosia Treitschke, 1833
Rhobonda Walker, 1864, preocc.
Carna Walker, 1864, repl. name
Gaeza Walker, 1864
Eurysara Turner, 1919
Euryzancla Turner, 1919
Macrozancla Turner, 1919
Brochometis Meyrick, 1923
Eunebristis Meyrick, 1923
Hapalonoma Meyrick, 1914
Ilingiotis Meyrick, 1914
Sirogenes Meyrick, 1923
Myrophila Meyrick, 1923
Onebala Walker, 1864
Helcystogramma Zeller, 1877
Dectobathra Meyrick, 1904
Pachysaris Meyrick, 1914
Paranoea Walsingham, 1911
Plocamosaris Meyrick, 1912
Noeza Walker, 1866, preocc.
Neochrista Meyrick, 1923
Prophoraula Meyrick, 1922
Semiomeris Meyrick, 1923
Teuchophanes Meyrick, 1914
Trichotaphe Clemens, 1860
Begoe Chambers, 1862
Malacotricha Zeller, 1873
Tricotaphe Riley, 1891, missp.
Malacotriche Busck, 1903, missp.
Malachotriche Busck, 1903, missp.
Vazugada Walker, 1864

Lecithocerinae

Deoclona Busck, 1903
Proclesis Walsingham, 1911
Lioclepta Meyrick, 1922
Deoclana Fletcher, 1929, missp.

Unplaced Genera

Acrophiletes Meyrick, 1932
Adullamitis Meyrick, 1932
Adullanitis Gaede, 1937, missp.
Aerotypia Walsingham, 1911
Agathactis Meyrick, 1929
Ageliarches Meyrick, 1923
Alsodryas Meyrick, 1914
Anomoxena Meyrick, 1917
Anthinora Meyrick, 1914
Anthistarcha Meyrick, 1925
Antistarcha Costa Lima, 1945, missp.
Apotactis Meyrick, 1918
Apothetoeca Meyrick, 1922
Arotromima Meyrick, 1929
Beltheca Busck, 1914
Anterethista Meyrick, 1914
Antherethista Gaede, 1937, missp.
Besciva Busck, 1914
Brachypsaltis Meyrick, 1931
Bruchiana Jörgensen, 1916
Catalexis Walsingham, 1909
Catoptristis Meyrick, 1925
Cerycangela Meyrick, 1925
Chalcomima Meyrick, 1929
Charistica Meyrick, 1925
Clistothyris Zeller, 1877
Coleostoma Meyrick, 1922
Colonanthes Meyrick, 1923

Commatica Meyrick, 1909
  Apopira Walsingham, 1911
Compsosaris Meyrick, 1914
  Gompsosaris Gaede, 1937, missp.
Copticostola Meyrick, 1929
Crambodoxa Meyrick, 1913
Darlia Clarke, 1950
Diastaltica Walsingham, 1910
Dissoptila Meyrick, 1914
Drepanoterma Walsingham, 1897
Elasiprora Meyrick, 1914
Empedaula Meyrick, 1918
Eripnura Meyrick, 1914
Eristhenodes Meyrick, 1935
Erythriastis Meyrick, 1925
Ethirostoma Meyrick, 1914
Eunomarcha Meyrick, 1923
  Atoponeura Busck, 1914, preocc.
Euzonomarcha Meyrick, 1925
Fortinea Busck, 1914
Galtica Busck, 1914
Glaucacna Forbes, 1931
  Glaucagna Gaede, 1937, missp.
Ilarches Meyrick, 1933
Iphimachaera Meyrick, 1931
Isembola Meyrick, 1926
Leistogenes Meyrick, 1927
Logisis Walsingham, 1909
Lophaeola Meyrick, 1932
Meridorma Meyrick, 1925
Metabolaea Meyrick, 1923
Metopleura Busck, 1912
Molopostola Meyrick, 1920
Oestomorpha Walsingham, 1911
Oxycryptis Meyrick, 1912
Oxylechia Meyrick, 1917
Pachygeneia Meyrick, 1923
Parastega Meyrick, 1912
Parelectroides Clarke, 1952, repl. name
  Parelectra Meyrick, 1925, preocc.
Pavolechia Busck, 1914
  Desmaucha Meyrick, 1918
Pelocnistis Meyrick, 1932
Perioristica Walsingham, 1910
Phylopatris Meyrick, 1923
Promolopica Meyrick, 1925
Ptilostonychia Walsingham, 1911
  Ptilonostychia Fletcher, 1929, missp.
Rhynchotona Meyrick, 1923
Satrapodoxa Meyrick, 1925
Sclerograptis Meyrick, 1923
Simoneura Walsingham, 1911
Scrotacta Meyrick, 1914
Stachyostoma Meyrick, 1923
Stagmaturgis Meyrick, 1923
Steremniodes Meyrick, 1923
Stereodmeta Meyrick, 1931
Stibarenches Meyrick, 1930
Symphanactis Meyrick, 1925
Synactias Meyrick, 1931
Tabernillaia Walsingham, 1911
  Tabernillaea Meyrick, 1925, emend.
Taphrosaris Meyrick, 1922
Tecia Kieffer & Jörgensen, 1910
  Fapua Kieffer & Jörgensen, 1910
  Lata Kieffer & Jörgensen, 1910
Trypsigenes Meyrick, 1914
  Thripsigenes Clarke, 1955, missp.
Thyrsomnestis Meyrick, 1929
Tocmia Walker, 1864
Trichembola Meyrick, 1918
Zelosyne Walsingham, 1911

Copromorphoidea

30. COPROMORPHIDAE

Cathelotis Meyrick, 1926
Endothamna Meyrick, 1922
Lotisma Busck, 1909
Neophylarcha Meyrick, 1926
Ordrupia Busck, 1911
  Ordupia Busck, 1911, missp.
  Ardrupia Busck, 1911, missp.
Phycomorpha Meyrick, 1914
Rhopalosetia Meyrick, 1926
Saridacma Meyrick, 1930
Syncamaris Meyrick, 1932

31. ALUCITIDAE

Alinguata Fleming, 1948
Alucita Linnaeus, 1758
  Orneodes Latreille, 1796, suppr.
  Euchiradia Hübner, [1825]
  Alucitina Heydenreich, 1851
Hexeretmis Meyrick, 1929
Paelia Walker, 1866

32. CARPOSINIDAE

Carposina Herrich-Schäffer, 1853
 Subgenus **Carposina** Herrich-Schäffer, 1853
 Subgenus **Trepsitypa** Meyrick, 1913
 Subgenus **Dipremna** Davis, 1969
 Subgenus **Epipremna** Davis, 1969
 Subgenus **Hypopremna** Davis, 1969
Atoposea Davis, 1969

33. EPERMENIIDAE

Ochromolopinae

Parochromolopis Gaedike, 1977

34. GLYPHIPTERIGIDAE

Cotaena Walker, [1865]
Myrsila Boisduval, [1875]
Phalerarcha Meyrick, 1913
Cronicombra Meyrick, 1920
Taeniostolella Fletcher, 1940, repl. name
  Taeniostola Meyrick, 1920, preocc.
Machlotica Meyrick, 1909
  Maclotica Busck, 1915, missp.
Neomachlotica Heppner, 1981
Trapeziophora Walsingham, 1892
Rhabdocrates Meyrick, 1931
Ussara Walker, 1864
  Setiostoma R. Felder & Rogenhofer, 1875
  Usara Busck, [1934], missp.
Sericostola Meyrick, 1927

Glyphipterix Hübner, [1825]
Heribeia Stephens, 1829
Aechmia Treitschke, 1833
Aecimia Boisduval, 1836, missp.
Glyphipteryx Zeller, 1839, emend.
Glyphiteryx Fischer von Röslerstamm, 1841, missp.
Anacampsoides Bruand, 1850, nom. oblit.
Glypipteryx Stainton, 1854, missp.
Glyphopteryx Herrich-Schäffer, 1854, emend.
Glyphiptoryx Mann & Rogenhofer, 1878, missp.
Glyphptieryx Turati, 1879, missp.
Glyphipterys Christoph, 1882, missp.
Glyphyteryx Hampson, 1918, missp.
Glyphteryx Watt, 1920, missp.
Diploschizia Heppner, 1981

Yponomeutoidea

35. PLUTELLIDAE

Plutellinae

Calliathla Meyrick, 1931
Eucalliathla Clarke, 1967
Euceratia Walsingham, 1881
Eudolichura Clarke, 1965
Leuroperna Clarke, 1965
Orthenches Meyrick, 1886
Philaustera Meyrick, 1927
Plutella Schrank, 1802
Anadetia Hübner, [1825]
Euota Hübner, [1825]
Cerostoma Stephens, 1834
Creagria Sodoffsky, 1837
Thalassonympha Meyrick, 1931
Ypsolopha Latreille, 1796
Ypsolophus Fabricius, 1798
Cerostoma Latreille, 1802
Hypsolopha Billberg, 1820
Theristis Hübner, [1825]
Harpipterix Hübner, [1825]
Abebaea Hübner, [1825]
Harpipteryx Treitschke, 1833, emend.
Chaetochilus Stephens, 1834
Harpepteryx Sodoffsky, 1837, emend.
Hypolepia Guenée, 1845, nom. nud.
Pteroxia Guenée, 1845, nom. nud.
Harpopteryx Agassiz, 1846, emend.
Hypsilophus Agassiz, 1846, emend.
Periclymenobius Wallengren, 1880
Credemnon Wallengren, 1880
Trachoma Wallengren, 1880
Pluteloptera Chambers, 1880
Plutelloptera Walsingham, 1881, missp.
Alapa Kieffer & Jörgensen, 1910
Mapa Strand, 1911
Pycnopogon Chrétien, 1922
Credemna Forbes, 1923, missp.
Melitonympha Meyrick, 1927
Chalconympha Meyrick, 1931
Credemon Moriuti, 1977, missp.

Praydinae

Atemelia Herrich-Schäffer, 1853
Prays Hübner, [1825]

36. YPONOMEUTIDAE

Attevinae

Atteva Walker, 1854
Peciloptera Clemens, 1860, preocc.
Amblothridia Wallengren, 1861
Corinea Walker, 1863
Oeta Grote, 1865
Carthara Walker, 1866, preocc.
Synadia Walker, 1866, repl. name
Scintilla Guenée, 1879, preocc.
Syblis Guenée, 1879
Lactura Walker, 1854
Dianasa Walker, 1854
Mieza Walker, 1854
Sarbena Walker, 1865, preocc.
Themiscyra Walker, 1865
Cyptasia Walker, 1866
Buxeta Walker, 1866
Enaemia Zeller, 1872
Pseudotalara Druce, 1885
Pseudocaprima Walsingham, 1900
Epidictica Turner, 1903
Hedycharis Turner, 1903
Eriopyrrha Meyrick, 1913
Pygmocrates Meyrick, 1932

Yponomeutinae

Anchimacheta Walsingham, 1914
Ditrigonophora Walsingham, 1897
Euarne Möschler, 1890
Ithutomus Butler, 1883
Ithytomus Meyrick, 1914, emend.
Spiladarcha Meyrick, 1913
Syncerastis Meyrick, 1931
Teinoptila Sauber, 1902
Toecorhychia Butler, 1883
Urodus Herrich-Schäffer, 1854
Trichostibas Zeller, 1863
Pexicnemidia Möschler, 1890, **n. syn.**
Paratiquadra Walsingham, 1897
Xyrosaris Meyrick, 1907
Xyrosaria Kearfott, [1903], missp.
Yponomeuta Latreille, 1796
Hyphantes Hübner, [1806], suppr.
Erminea Haworth, [1811]
Hyponomeuta Billberg, 1820, emend.
Coenyphantes Hübner, [1822]
Nygmia Hübner, [1825], preocc.
Hyponomeuta Sodoffsky, 1837, emend.
Hyponomenta Turner, 1898, missp.
Zelleria Stainton, 1849

37. ARGYRESTHIIDAE

Argyresthia Hübner, [1825]
Argyrosetia Stephens, 1829
Oligos Treitschke, 1830
Ederesa Curtis, 1833
Ismene Stephens, 1834

Blastotere Ratzeburg, 1840
Argyrestia MacKay, 1972, missp.

38. DOUGLASIIDAE

Protonyctia Meyrick, 1931

39. ACROLEPIIDAE

Acrolepiopsis Gaedike, 1970
Argiope Chambers, 1873, preocc.
Antispastis Meyrick, 1926

40. HELIODINIDAE

Schreckensteiniinae

Schreckensteinia Hübner, [1825]
Chrysocorys Curtis, 1833

Heliodininae

Amphiclada Meyrick, 1912
Copocentra Meyrick, 1909
Crembalastis Meyrick, 1915
Cycloplastis Clemens, 1864
Heliodines Stainton, 1854
Aetole Chambers, 1875
Aetola Frey, 1884, missp.
Heliodinides Turner, 1941, missp.
Lamprolophus Busck, 1900
Embola Walsingham, 1909
Lithariapteryx Chambers, 1876
Pseudastasia Walsingham, 1909
Scelorthus Busck, 1900
Thrasydoxa Meyrick, 1912

Immoidea

41. IMMIDAE

Moca Walker, 1863
Adricara Walker, 1863
Alicadra Walker, [1866]
Jobula Walker, 1866
Callartona Hampson, [1893]
Loxotrochis Meyrick, 1906
Imma Walker, [1859]
Pingrassa Walker, [1859]
Tortricomorpha C. Felder, 1861
Topaza Walker, 1864
Vinzela Walker, [1866]
Thylacopleura Meyrick, 1886
Davendra Moore, 1887
Pseudotortrix Turner, 1900

LEPIDOPTERA

Suborder Zeugloptera

Micropterigoidea

1. MICROPTERIGIDAE

by N. P. Kristensen & E. S. Nielsen

HYPOMARTYRIA Kristensen & Nielsen, 1982

1	**micropteroides** Kristensen & Nielsen, 1982	Chile

SQUAMICORNIA Kristensen & Nielsen, 1982

2	**aequatoriella** Kristensen & Nielsen, 1982	Ecuador

Suborder Heterobathmiina

Heterobathmioidea

2. HETEROBATHMIIDAE

by N. P. Kristensen & E. S. Nielsen

HETEROBATHMIA Kristensen & Nielsen, 1979

1	**diffusa** Kristensen & Nielsen, 1979	Argentina
2	**pseuderiocrania** Kristensen & Nielsen, 1979	Argentina

Suborder Glossata

Infraorder Neopseustina

Neopseustoidea

3. NEOPSEUSTIDAE

by D. R. Davis

SYNEMPORA Davis & Nielsen, 1980

1	**andesae** Davis & Nielsen, 1980	Argentina

APOPLANIA Davis, 1975

2	**penai** Davis & Nielsen, 1980	Chile
3	**valdiviae** Davis & Nielsen, 1984	Chile
4	**chilensis** Davis, 1975	Chile

Infraorder Exoporia

Hepialoidea

4. NEOTHEORIDAE

by N. P. Kristensen

NEOTHEORA Kristensen, 1978

1	**chiloides** Kristensen, 1978	Brazil (MT)

5. HEPIALIDAE

by G. S. Robinson & E. S. Nielsen

DALACA Walker, 1856
Dolaca Druce, 1901, missp.
Huapina Bryk, 1945
Maculella Viette, 1950
Toenga Tindale, 1954

1	**crocatus** (Ureta, 1956) (Hepialus)	Chile
2	**chiliensis** (Viette, 1950) (Maculella)	Chile
	chilensis (Viette, 1950) (Maculella), missp.	
3	**pallens** (Blanchard, 1852) (Hepialus)	Chile
	hemileuca Butler, 1882	Chile
	marmorata Butler, 1882	Chile
	subfervens Butler, 1882	Chile
	violacea Butler, 1882	Chile
	dimidiatus (Berg, 1882) (Aepytus)	Chile
	noctuides Pfitzner, 1914	Chile
	parviguttata (Bryk, 1945) (Huapina)	Argentina
	pseudodimiata (Paclt, 1953) (Lossbergiana)	Argentina
	oceanica (Tindale, 1954) (Toenga)	"Cook Is." [Chile?]
4	**quadricornis** Nielsen & Robinson, 1983	Argentina
5	**nigricornis** Walker, 1856	Chile
6	**patriciae** Nielsen & Robinson, 1983	Argentina
7	**laminata** Nielsen & Robinson, 1983	Chile
8	**fuscus** (Mabille, 1885) (Hepialus)	Chile
9	**postvariabilis** Nielsen & Robinson, 1983	Argentina
10	**variabilis** (Viette, 1950) (Maculella)	Chile

CALLIPIELUS Butler, 1882
Stachyocera Ureta, 1957

11	**arenosus** Butler, 1882	Chile
	antarcticus (Staudinger, 1899) (Hepialus), preocc. (not Wallengren, 1860)	Argentina
	staudingeri Wagner, 1911, repl. name	Argentina
	leukogramma Bryk, 1945	Argentina
	chiliensis Viette, 1950	Chile
12	**digitata** Robinson, 1977	Chile
	brunnescens Robinson, 1977	Chile
	castilloi Robinson, 1977	Chile
13	**salasi** Robinson, 1977	Chile
14	**perforata** Nielsen & Robinson, 1983	Argentina
15	**gentilii** Nielsen & Robinson, 1983	Argentina
16	**fumosa** Nielsen & Robinson, 1983	Chile
17	**argentata** Ureta, 1957	Chile
18	**krahmeri** Nielsen & Robinson, 1983	Chile
19	**izquierdoi** (Ureta, 1957) (Stachyocera)	Chile
20	**vulgaris** Nielsen & Robinson, 1983	Argentina

BLANCHARDINA Viette, 1950

21	**venosus** (Blanchard, 1852) (Hepialus)	Chile

CALADA Nielsen & Robinson, 1983

22	**fuegensis** Nielsen & Robinson, 1983	Argentina
23	**migueli** Nielsen & Robinson, 1983	Argentina

PUERMYTRANS Viette, 1951

24	**chiliensis** Viette, 1951	Chile

PARAPIELUS Viette, 1949
Lossbergiana Viette, 1950

25	**luteicornis** (Berg, 1882) (Pielus)	Argentina
	f. popperi (Pfitzner, 1938) (Pielus)	[Argentina?]
26	**oberthuri** (Viette, 1950) (Lossbergiana)	Chile
27	**heimlichi** (Ureta, 1956) (Hepialus)	Chile
28	**reedi** (Ureta, 1957) (Hepialus)	Chile

ANDEABATIS Nielsen & Robinson, 1983

29	**chilensis** (Ureta, 1951) (Xyleutes)	Chile

DRUCEIELLA Viette, 1949

30	**basirubra** (Schaus, 1901) (Dalaca)	Peru
	songoensis (Pfitzner, 1914) (Pseudophassus)	Bolivia
31	**amazonensis** Viette, 1950	Brazil (Pa)
32	**metellus** (Druce, 1890) (Hepialus)	Ecuador
33	**momus** (Druce, 1890) (Hepialus)	Ecuador
	metricus (Pfitzner, 1914) (Pseudophassus), nom. nud.	?
	f. metricus (Pfitzner, 1938) (Pseudophassus)	?

TRICHOPHASSUS Le Cerf, 1919

34	**giganteus** (Herrich-Schäffer, [1853]) (Epiolus [sic])	"S. Am."
	hayeki (Foetterle, 1903) (Phassus)	Brazil (RJ)

PHASSUS Walker, 1856

35	**triangularis** H. Edwards, 1885	Mexico (Ver)
	f. triangularides Pfitzner, 1938	Mexico
36	**huebneri** (Geyer, [1838]) (Pharmacis)	Mexico
	argentiferus Walker, 1856	Mexico (Oax)
	pedipogon Strand, 1916	Costa Rica
37	**basirei** Schaus, 1890	Mexico (Ver)
38	**n-signatus** Weymer, 1907	Guatemala
39	**phalerus** Druce, 1887	Mexico (Ver)
40	**marcius** Druce, 1892	Mexico (Dur)
41	**exclamationis** Pfitzner, 1938	?
42	**aurigenus** Pfitzner, 1914	Costa Rica
43	**championi** Druce, 1887	Guatemala
44	**pharus** (Druce, 1887) (Hepialus)	Guatemala
45	**rosulentus** Weymer, 1907	Mexico (Ver)
46	**eldorado** Pfitzner, 1906	Venezuela
47	**pretiosus** (Herrich-Schäffer, [1856]) (Epialus [sic])	Brazil
	plusia (Herrich-Schäffer, [1856]) (Epialus [sic]), unavail.	Brazil
48	**agrionides** Walker, 1856	Brazil
49	**tessellatus** (Herrich-Schäffer, [1854]) (Epialus [sic])	"New Holland"
50	**smithi** Druce, 1889	Mexico (Ver)
51	**costaricensis** Druce, 1887	Costa Rica
52	**absyrtus** Schaus, 1892	Brazil (RJ)
53	**guianensis** Schaus, 1940	Guyana

54	**chrysodidyma** Dyar, 1915	Mexico (Mex)
55	**transversus** Walker, 1856	Brazil (RJ)

SCHAUSIANA Viette, 1949

56	**trojesa** (Schaus, 1901) (Phassus)	Mexico (Jal)

APLATISSA Viette, 1953

57	**michaelis** (Pfitzner, 1914) (Dalaca)	Brazil (Am)/Peru
	michaeli (Pfitzner, 1937) (Dalaca), missp.	
58	**strangoides** Viette, 1953	Brazil (Am)

PFITZNERIANA Viette, 1952
Pseudophassus Pfitzner, 1938, preocc. (Pfitzner, 1914 [Hepialidae])

59	**olivescens** (Pfitzner, 1914) (Dalaca)	
	a) **olivescens** (Pfitzner, 1914) (Dalaca)	Colombia
	b) **boliviensis** Viette, 1961	Bolivia
60	**vogli** Viette, 1952	Venezuela
61	**allura** Viette, 1961	Bolivia
62	**prosopus** (Druce, 1901) (Hepialus)	Colombia

AEPYTUS Herrich-Schäffer, [1858]

Subgenus **Hampsoniella** Viette, 1949

63	**equatorialis** Viette, 1949	Ecuador
64	**serta** (Schaus, 1894) (Dalaca)	Mexico (Ver)
65	**assa** (Druce, 1887) (Dalaca)	Guatemala

Subgenus **Pseudodalaca** Viette, 1949

66	**mexicanensis** (Viette, 1952) (Pseudodalaca)	Mexico (Ver)
67	**gugelmanni** Viette, 1949	Mexico (Ver)

Subgenus **Gymelloxes** Viette, 1952

68	**terea** (Schaus, 1892) (Dalaca)	Mexico ("Paso San Juan")
	muysca (Pfitzner, 1914) (Dalaca)	Panama
69	**trilinearis** (Pfitzner, 1914) (Dalaca)	Colombia
	trilinearides (Pfitzner, 1937) (Dalaca), missp.	
70	**paropus** (Druce, 1890) (Hepialus)	Ecuador

Subgenus **Alloaepytus** Viette, 1952

71	**tesselloides** (Schaus, 1901) (Dalaca)	Paraguay
	coscinophora (Pfitzner, 1914) (Dalaca)	Brazil (MT)

Subgenus **Aepytus** Herrich-Schäffer, [1858]

72	**jeanneli** (Viette, 1950) (Schaefferiana)	Brazil (Pr)
73	**biedermanni** (Viette, 1950) (Schaefferiana)	Brazil (MG)
74	**exclamans** (Herrich-Schäffer, [1854]) (Epialus [sic])	Brazil
75	**forsteri** Viette, 1961	Bolivia
76	**munona** Schaus, 1929	Brazil (SC)
77	**petropolisiensis** Viette, 1951	Brazil (RJ)
78	**helga** Schaus, 1929	Brazil (SC)
79	**zischkai** Viette, 1961	Bolivia
80	**danieli** Viette, 1961	Argentina

Subgenus **Thiastyx** Viette, 1951

81	**catharinae** (Viette, 1951) (Thiastyx)	Brazil (SC)

Subgenus **Schaefferiana** Viette, 1950

82	**epigramma** (Herrich-Schäffer, [1854]) (Epialus [sic])	Brazil
83	**simplex** (Viette, 1955) (Schaefferiana)	Brazil (MG)

Subgenus **Paragorgopis** Viette, 1951

84	**foetterlei** (Viette, 1951) (Paragorgopis)	Brazil (RJ)
85	**oreas** (Schaus, 1892) (Dalaca)	Brazil (RJ)
86	**spitzi** (Viette, 1955) (Paragorgopis)	Brazil (SP)
87	**jordani** (Viette, 1955) (Paragorgopis)	Brazil (MG)
88	**pittionii** (Viette, 1951) (Paragorgopis)	Brazil (RJ)
89	**schausi** (Viette, 1951) (Paragorgopis)	Brazil (SP)
90	**nigrovenosalis** (Viette, 1955) (Paragorgopis)	Brazil (MG)

Subgenus **Hepialyxodes** Viette, 1951

91	**rileyi** (Viette, 1951) (Hepialyxodes)	Brazil (SP)

Subgenus **Xytrops** Viette, 1950

92	**dorita** (Schaus, 1901) (Cibyra)	Brazil (Pr)
93	**monoargenteus** Viette, 1950	Brazil (Pr)
94	**yungas** (Viette, 1961) (Xytrops)	Bolivia
95	**pluriargenteus** (Viette, 1955) (Xytrops)	Brazil (SP)
96	**verresi** Schaus, 1929	Brazil (SC)

Subgenus **Cibyra** Walker, 1856

97	**poltrona** (Schaus, 1901) (Cibyra)	Brazil (Pr)
98	**ferruginosa** (Walker, 1856) (Cibyra)	Brazil
	ferruginea (Kirby, 1892) (Cibyra), missp.	
	dormita (Schaus, 1901) (Cibyra)	Brazil (RJ)

Subgenus **Lamelliformia** Viette, 1951

99	**sladeni** (Hampson, 1903) (Dalaca)	Brazil (MT)
	tupi (Pfitzner, 1914) (Cibyra)	Brazil (RJ)
100	**prytanes** (Schaus, 1892) (Dalaca)	Brazil (RJ)

Subgenus **Tricladia** R. Felder, 1874
Pseudophassus Pfitzner, 1914
Parana Viette, 1949

101	**mahagoniatus** (Pfitzner, 1914) (Pseudophassus)	Bolivia
102	**umbrifera** (R. Felder, 1874) (Tricladia)	Brazil
103	**philiponi** Viette, 1949	Brazil (Pa)

Subgenus **Pseudophilaenia** Viette, 1950

104	**omagua** (Pfitzner, 1937) (Philaenia)	Peru

Subgenus **Philaenia** Kirby, 1892

105	**guyanensis** Viette, 1950	French Guiana
106	**lagopus** (Möschler, 1877) (Pharmacis)	Surinam
107	**thisbe** (Druce, 1901) (Dolaca [sic])	Colombia
	f. hemichrysea (Pfitzner, 1937) (Dalaca)	Colombia
108	**indicata** (Strand, 1912) (Dalaca)	Ecuador
109	**brasiliensis** (Viette, 1951) (Philaenia)	Brazil (RJ)
110	**saguanmachica** (Pfitzner, 1914) (Dalaca)	Colombia
111	**fasslii** (Pfitzner, 1914) (Dalaca)	Colombia

Subgenus **Yleuxas** Viette, 1951

112	**brunnea** (Schaus, 1901) (Cibyra)	Venezuela
	bradleyi (Viette, 1951) (Yleuxas)	Peru

PHIALUSE Viette, 1961

113	**palmar** Viette, 1961	Bolivia

ROSEALA Viette, 1950

114	**bourgognei** Viette, 1950	Brazil (RJ)

"DALACA" auct., not Walker, 1856 [1]

115	**chiriquensis** Pfitzner, 1914	Panama
116	**cocama** Pfitzner, 1914	Peru
	nannophyes Pfitzner, 1914	Ecuador
117	**cuprifera** Pfitzner, 1914	Peru
118	**guarani** Pfitzner, 1914	Brazil (SC)
119	**katharinae** Pfitzner, 1914	Brazil (SC)
120	**manoa** Pfitzner, 1914	Colombia
121	**niepelti** Pfitzner, 1914	Ecuador
122	**obliquestrigata** Strand, 1912	Peru
123	**stigmatica** Pfitzner, 1937	Paraguay
124	**tapuja** Pfitzner, 1914	Colombia
125	**usaque** Pfitzner, 1914	Colombia
126	**vibicata** Pfitzner, 1914	Ecuador
127	**mummia** Schaus, 1892	Brazil (RJ)
	mummea Pfitzner, 1937, missp.	

PFITZNERIELLA Viette, 1950

128	**lucicola** (Maassen, 1890) (Triodia)	Ecuador
129	**monticola** (Maassen, 1890) (Triodia)	Ecuador
130	**similis** (Zukowsky, 1954) (Triodia)	Peru
131	**remota** (Pfitzner, 1906) (Hepialus)	Peru

6. PALAEOSETIDAE [2]

by G. S. Robinson & E. S. Nielsen

OSRHOES Druce, 1900

1	**coronta** Druce, 1900	Colombia

Infraorder Heteroneura

Division Monotrysia

Nepticuloidea

7. NEPTICULIDAE

by D. R. Davis

Nepticulinae [3]

Tribe Nepticulini [4]

STIGMELLA Schrank, 1802
Nepticula Heyden, 1843
Stigmalla Herrich-Schäffer, 1853, missp.
Dysnepticula Börner, 1925
Johanssonia Borkowski, 1972

1	**aerifica** (Meyrick, 1915) (Nepticula), **n. comb.**	Peru
2	**andina** (Meyrick, 1915) (Nepticula), **n. comb.**	Peru
3	**costalimai** (Bourquin, 1962) (Nepticula), **n. comb.**	Argentina
4	**cuprata** (Meyrick, 1915) (Nepticula), **n. comb.**	Peru
5	**epicosma** (Meyrick, 1915) (Nepticula), **n. comb.**	Peru
6	**eurydesma** (Meyrick, 1915) (Nepticula), **n. comb.**	Guyana
7	**gossypii** (Forbes, 1930) (Nepticula), **n. comb.**	Puerto Rico
8	**guittonae** (Bourquin, 1962) (Nepticula), **n. comb.**	Argentina
9	**hylomaga** (Meyrick, 1931) (Nepticula), **n. comb.**	Argentina
10	**johannis** (Zeller, 1877) (Nepticula), **n. comb.**	Colombia
11	**molybditis** (Zeller, 1877) (Nepticula), **n. comb.**	Colombia
12	**olyritis** (Meyrick, 1915) (Nepticula), **n. comb.**	Peru

Tribe Trifurculini

ENTEUCHA Meyrick, 1915

13	**cyanochlora** Meyrick, 1915	Guyana

8. OPOSTEGIDAE

by D. R. Davis

OPOSTEGA Zeller, 1839

1	**abrupta** Walsingham, 1897	Virgin Is. (St. Thomas)
2	**accessoriella** Frey & Boll, 1876	USA (Tx)
3	**acidata** Meyrick, 1915	Ecuador
4	**adusta** Walsingham, 1897	Virgin Is. (St. Thomas)
5	**congruens** Walsingham, 1914	Mexico (Gue)
6	**elachista** Walsingham, 1914	Mexico (Gue)
7	**microlepta** Meyrick, 1915	Ecuador/Guyana
8	**monosperma** Meyrick, 1931	Brazil (MG)
9	**paromias** Meyrick, 1915	Peru
10	**perdigna** Walsingham, 1914	Mexico (Gue)
11	**pexa** Meyrick, 1920	Brazil (Pa)
12	**pontifex** Meyrick, 1915	Colombia
13	**protomochla** Meyrick, 1935	Argentina
14	**pumila** Walsingham, 1914	Mexico (Tab)
15	**sacculata** Meyrick, 1915	Ecuador
16	**saltatrix** Walsingham, 1897	Virgin Is. (St. Thomas)

9. TISCHERIIDAE

by D. R. Davis

TISCHERIA Zeller, 1839
Evexia Gistel, 1847
Philodoxa Gistel, 1848
Coptotriche Walsingham, 1890
Tisheria Busck, [1903], missp.

1	**capnota** Meyrick, 1915	Peru
2	**deliquescens** Meyrick, 1915	Guyana
3	**elongata** Walsingham, 1914	Mexico (Gue)
4	**ephaptis** Meyrick, 1915	Peru
5	**koehleri** Bourquin, 1962	Argentina
6	**plagifera** Meyrick, 1915	Ecuador
7	**pulverea** Walsingham, 1897	Virgin Is. (St. Thomas)
8	**unicolor** Walsingham, 1897	Virgin Is. (St. Croix)

Incurvarioidea

10. INCURVARIIDAE

by D. R. Davis

BASILEURA Nielsen & Davis, 1981

1	**elongata** Nielsen & Davis, 1981	Argentina

SIMACAUDA Nielsen & Davis, 1981

2	**dicommatias** (Meyrick, 1931) (Lampronia)	Chile
3	**heliocephala** (Meyrick, 1931) (Lampronia)	Chile
4	**virescens** Nielsen & Davis, 1981	Argentina

11. CECIDOSIDAE

by D. R. Davis

CECIDOSES Curtis, 1835
Clistoses Kieffer & Jörgensen, 1910
Eucecidoses Brèthes, 1916

1	**eremita** Curtis, 1835	Uruguay
	artifex (Kieffer & Jörgensen, 1910) (Clistoses)	Argentina
2	**minutanus** Brèthes, 1916	Argentina

OLIERA Brèthes, 1916

3	**argentinana** Brèthes, 1916	Argentina

RIDIASCHINIA Brèthes, 1916

4	**congregatella** Brèthes, 1916	Argentina

DICRANOSES Kieffer & Jörgensen, 1910

5	**capsulifex** Kieffer & Jörgensen, 1910	Argentina

12. ADELIDAE

by D. R. Davis

CEROMITIA Zeller, 1854
Trichorrhabda Meyrick, 1912, **n. syn.**

1	**chionocrossa** Meyrick, 1921	Brazil (Pa)
2	**eremarcha** Meyrick, 1914	Paraguay
3	**exalbata** Meyrick, 1921	Brazil (Pa)
4	**fasciolata** (Butler, 1883) (Nemophora)	Chile
5	**ilyodes** Meyrick, 1931	Argentina
6	**laminensis** Pastrana, 1961	Argentina
7	**lizeri** Pastrana, 1961	Argentina
8	**ochrodyta** Meyrick, 1921	Brazil (Am)
9	**phaeoceros** Meyrick, 1921	Brazil (Am)
10	**pucaraensis** Pastrana, 1961	Argentina
11	**schajovskoii** Pastrana, 1961	Argentina
12	**sciographa** Meyrick, 1921	Brazil (Pa)
13	**viscida** Meyrick, 1921	Brazil (Pa)

ADELA Latreille, 1796
Capillaria Haworth, 1828
Metallitis Sodoffsky, 1837
Aedilis Gistel, 1873
Dicte Chambers, 1873

14	**aethiops** R. Felder & Rogenhofer, 1875	"Australia" [C. Am.]
15	**astrella** Walsingham, 1915	Mexico (Gue)
16	**lithopola** Walsingham, 1915	Mexico (Ver)
17	**ridingsella** Clemens, 1864	USA (Va)

13. PRODOXIDAE

by D. R. Davis

TEGETICULA Zeller, 1873
Pronuba Riley, 1872, preocc. (Thompson, 1860 [Coleoptera])
Promiba Kirby, 1874, missp.
Thia H. Edwards, 1888, preocc. (Leach, 1815 [Crustacea])
Thelethia Dyar, 1893
Valentinia Coolidge, 1909, preocc. (Walsingham, 1907 [Blastobasidae])

*1	**maculata** (Riley, 1881) (Pronuba)	
	a) **extranea** (H. Edwards, 1888) (Thia)	USA (Ca)
	aterrima (Trelease, 1893) (Pronuba)	USA (Ca)
2	**yuccasella** (Riley, 1872) (Pronuba)	USA (Mo)
	alba Zeller, 1873	USA (Tx)
	yuccaella (Boll, 1876) (Pronuba), missp.	
	intermedius (Riley, 1881) (Prodoxus)	USA (Tx)
	mexicana Bastida, 1962	Mexico (SLP)

PARATEGETICULA Davis, 1967

3	**pollenifera** Davis, 1967	USA (Az)

PRODOXUS Riley, 1880

4	**quinquepunctellus** (Chambers, 1875) (Hyponomeuta [sic])	USA (Tx)
	paradoxica (Chambers, 1878) (Hyponomeuta [sic])	USA (Co)
	decipiens Riley, 1880	USA (SC)
5	**intricatus** Riley, 1893	Mexico (Ver)

PRODOXOIDES Nielsen & Davis, 1984 [5]

6	**assymmetra** Nielsen & Davis, 1984	Chile

14. HELIOZELIDAE

by J. B. Heppner

ANTISPILA Hübner, [1825]

1	**chlorosema** Meyrick, 1931	Chile
2	**cyclosema** Meyrick, 1921	Brazil (Am)
3	**mesogramma** Meyrick, 1921	Peru
3.1	**nolckeni** Zeller, 1877	Colombia
4	**orthodelta** Meyrick, 1931	Brazil (Am)
5	**pentalitha** Meyrick, 1916	Guyana
6	**postscripta** Meyrick, 1921	Peru
7	**praecincta** Meyrick, 1921	Brazil (Pa)

8 **trypherantis** Meyrick, 1916 Guyana

COPTODISCA Walsingham, 1895
Aspidisca Clemens, 1860, preocc. (Ehrenburg, 1830 [Protura])

9 **rhizophorae** Walsingham, 1897 Virgin Is. (St. Thomas)

HELIOZELA Herrich-Schäffer, 1853
Dyselachista Spuler, 1910

10 **ahenea** Walsingham, 1897 Haiti
11 **cuprea** Walsingham, 1897 Virgin Is. (St. Thomas)

LAMPROZELA Meyrick, 1916

12 **praefulgens** Meyrick, 1916 Guyana

MONACHOZELA Meyrick, 1931

13 **neoleuca** Meyrick, 1931 Brazil (Pa)

PHANEROZELA Meyrick, 1921

14 **polydora** Meyrick, 1921 Brazil (Am)

Division Ditrysia

Section Tineina

Tineoidea

15. TINEIDAE [6]

by D. R. Davis

ACROLOPHUS Poey, 1832
Anaphora Clemens, 1859
Daulia Walker, 1863, preocc. (Walker, 1859 [Pyralidae])
Derchis Walker, 1863
Eddara Walker, 1863, preocc. (Walker, 1858 [Hemiptera])
Hibita Walker, 1863
Tirasia Walker, 1863
Bazira Walker, 1864
Phlongia Walker, 1864
Urbara Walker, 1864
Tachasara Walker, 1865
Eutheca Grote, 1881, preocc. (Kiesenwetter, 1877 [Coleoptera])
Eulepiste Walsingham, 1882
Pseudoconchylis Walsingham, 1884
Homonymus Walsingham, 1887
Neolophus Walsingham, 1887
Hypoclopus Walsingham, 1887
Thysanoscelis Walsingham, 1887
Thysanoskelis Walsingham, 1887, missp.
Ankistrophorus Walsingham, 1887
Caenogenes Walsingham, 1887
Felderia Walsingham, 1887
Ortholophus Walsingham, 1887
Pseudanaphora Walsingham, 1887
Sapinella Kirby, 1892
Atopocera Walsingham, 1897
Pilanaphora Walsingham, 1897
Thysanosedes Druce, 1901, missp.
Hypocolypus Dyar, [1903], missp.
Anophora Kearfott, 1903, missp.
Brachysymbola Meyrick, 1912
Orothyntis Meyrick, 1913
Apoclisis Walsingham, 1914
Psephocrita Meyrick, 1919
Xerocausta Meyrick, 1928, **n. syn.**
Anaphorina Strand, 1932

1 **abdita** Walsingham, 1914 Guatemala
*2 **acornus** Hasbrouck, 1964 USA (Az)
3 **albipennis** Meyrick, 1931 Brazil (MG?)
4 **anaphorella** (Walsingham, 1892) (Amydria), **n. comb.** St. Vincent
5 **anathyrsa** Meyrick, 1919 French Guiana
6 **angulatella** (Walsingham, 1897) (Morophaga), **n. comb.** Dominica
7 **apertella** Busck, 1913 Mexico (Mex)
8 **arcasalis** (Walker, 1858) (Palthis) Dom. Rep.
9 **arcturella** (Walker, 1863) (Hibita) Brazil (Am)
10 **argentinus** Walsingham, 1887 Argentina
11 **arida** Walsingham, 1914 Mexico (Gue)
12 **arimasalis** (Walker, 1850) (Palthis) Brazil (Pa)
*13 **arizonellus** Walsingham, 1887 USA (Az)
14 **australis** (Walsingham, 1897) (Xylesthia), **n. comb.** Haiti
15 **bactra** Busck, 1914 Panama
*16 **baldufi** Hasbrouck, 1964 USA (Az)
17 **barbipalpus** Busck, 1912 Mexico (Pue)
18 **barema** Durrant, 1915 Guatemala
19 **baryspila** Meyrick, 1931 Brazil (ES)
20 **bidens** Walsingham, 1914 Mexico (Dur)
21 **bifurcata** Busck, 1914 Panama
22 **bogotensis** (Walsingham, 1887) (Anaphora) Colombia
23 **bombaulia** Meyrick, 1922 Argentina
24 **boucardi** Druce, 1901 Mexico (Ver)
25 **bugabae** Walsingham, 1914 Panama
26 **capax** Meyrick, 1928 Colombia
27 **catagnampta** Meyrick, 1930 Brazil (Pa)
28 **ceramocha** (Meyrick, 1928) (Xerocausta), **n. comb.** Colombia
29 **cervicolor** Meyrick, 1931 Argentina
30 **chloropelta** Meyrick, 1922 Brazil (Am)
31 **chonactis** Meyrick, 1932 Brazil (RJ)
32 **cleptica** Walsingham, 1914 Mexico (Gue)
*33 **cockerelli** (Dyar, 1900) (Eulepiste) USA (NM)
34 **condita** Durrant, 1914 Mexico (Ver)
35 **contubernalis** Meyrick, 1922 Brazil (RJ)
36 **corrientis** (Walsingham, 1887) (Ankistrophorus) Argentina
37 **corticinicolor** Strand, 1920 Costa Rica
38 **corvula** Walsingham, 1914 Mexico (Dur)
39 **corymba** Durrant, 1914 Guatemala
40 **cosmeta** Walsingham, 1914 Guatemala
41 **cossoides** R. Felder & Rogenhofer, 1875 Brazil
42 **crinifrons** Walsingham, 1914 Mexico (Ver)
43 **cyclophora** Meyrick, 1931 Paraguay
44 **damina** Walsingham, 1914 Panama
45 **diachelota** Meyrick, 1931 Argentina
46 **dictyopsamma** Meyrick, 1931 Brazil (ES)
47 **dimidiella** (Walsingham, 1892) (Felderia) Cuba
48 **directus** Busck, 1912 Mexico (DF)
49 **doeri** (Walsingham, 1887) (Felderia) Brazil (RJ)
50 **ductifera** Meyrick, 1927 Antigua
51 **echinon** (Druce, 1901) (Felderia) Mexico (Ver)
cassicordis Dyar, 1907 Mexico (DF)
robertus Busck, 1920, **n. syn.** Mexico (Mex)
52 **echinura** Meyrick, 1915 Peru
53 **ectenes** Walsingham, 1914 Mexico (Gue)
54 **empedocles** Meyrick, 1930 Brazil (RJ)
55 **emphytopa** Meyrick, 1932 Brazil (RS)
56 **erethismia** Meyrick, 1930 Brazil (Pa)
57 **euporia** Walsingham, 1914 Mexico (Gue)
58 **euteles** Walsingham, 1914 Mexico (Ver)
59 **exigua** Meyrick, 1915 Guyana
60 **farracea** Meyrick, 1931 Brazil (ES)
61 **ferrarenella** (Walker, 1864) (Phlongia) [S. Am.?]
62 **ferruginea** (Walsingham, 1887) (Anaphora) Colombia
63 **fervidus** Busck, 1912 Mexico (Ver)
antonellus (Barnes & McDunnough, 1913) (Neolophus) USA (Tx)
64 **filicicornis** (Walsingham, 1887) (Felderia) USA (Az)
mexicanellus Beutenmüller, 1888 Mexico (DF)
filicornis (Dyar, 1900) (Felderia), missp.
65 **forreri** Walsingham, 1914 Mexico (Dur)
66 **fumida** Walsingham, 1914 Costa Rica
67 **galeata** (Walker, 1864) (Urbara) Brazil (Am)
68 **garleppi** (Druce, 1901) (Felderia) Bolivia
69 **giganteus** (Druce, 1901) (Ankistrophorus) Mexico (Ver)
70 **goniocentra** Meyrick, 1923 Brazil (Am)
*71 **griseus** (Walsingham, 1887) (Hypoclopus)
a) **griseus** (Walsingham, 1887) (Hypoclopus) USA (Az)
leucallactis Meyrick, 1919 USA (Az)
72 **halidora** Meyrick, 1915 Guyana
73 **hamiferella** (Hübner, [1827-31]) (Pinaris) Brazil (RJ)
74 **harmoniella** Busck, 1913 Mexico (Pue)
75 **harparsen** Forbes, 1931 Puerto Rico
76 **hedemanni** (Walsingham, 1897) (Pilanaphora) Virgin Is. (St. Croix)
77 **hirsutevestita** (Walsingham, 1897) (Morophaga) Jamaica
78 **hirsutus** (Walsingham, 1887) (Thysanoscelis) Brazil (ES)
79 **horridalis** (Walker, 1863) (Derchis) Brazil (Am)
80 **hypophaea** Meyrick, 1923 Brazil (Am)
81 **icarus** Busck, 1912 Mexico (Pue)
82 **illudens** Meyrick, 1924 Jamaica
83 **infida** Meyrick, 1912 Venezuela
84 **interfusa** Meyrick, 1927 Ecuador
85 **invida** Durrant, 1914 Panama
86 **irrisoria** Meyrick, 1924 Jamaica
87 **jalapae** Walsingham, 1914 Mexico (Ver)
harpella Walsingham, 1914, **n. syn.** Mexico (Ver)
*88 **kearfotti** (Dyar, 1903) (Eulepiste) USA (Az)
diversus Busck, 1912 USA (Ca)
*88.1 **klotsi** Hasbrouck, 1964 USA (Az)
89 **laetifica** Durrant, 1914 Mexico (Ver)
90 **latiberbis** Meyrick, 1931 Brazil (SP)
91 **lerodes** Durrant, 1914 Mexico (Gue)
92 **leucodocis** (Zeller, 1877) (Anaphora) Cuba
medioliniella (Kearfott, 1907) (Dorata) USA (Ca)
93 **leucopogon** Walsingham, 1914 Mexico (Tab)
94 **leucotricha** Meyrick, 1931 Paraguay
95 **libitina** Druce, 1901 Guatemala
96 **linus** Druce, 1901 Mexico (Ver)
97 **lithopa** Durrant, 1914 Mexico (Dur)
*98 **macrogaster** (Walsingham, 1887) (Anaphora) USA (Az)
bipectinicornis Hasbrouck, 1964, **n. syn.** [7] USA (Az)
unipectinicornis Hasbrouck, 1964, **n. syn.** USA (Az)
laminicornis Hasbrouck, 1964, **n. syn.** USA (Az)
*99 **macrophallus** Hasbrouck, 1964 USA (Az)

100 **macrozancla** Meyrick, 1922 Peru
101 **maculata** Walsingham, 1887 Brazil (RJ)
102 **maculisecta** Busck, 1913 Mexico (Ver)
dentiger Walsingham, 1914, **n. syn.** Mexico (Ver)
103 **manticodes** Meyrick, 1919 French Guiana
104 **marcida** Walsingham, 1914 Mexico (Gue)
105 **merocoma** Meyrick, 1922 Peru
106 **micromacha** Meyrick, 1932 Costa Rica
107 **mimasalis** (Walker, 1858) (Palthis) Dom. Rep.
indecora (Walker, 1863) (Daulia), **n. syn.** Dom. Rep.
languidalis (Walker, 1865) (Tachasara), **n. syn.** Dom. Rep.
108 **minima** (Walsingham, 1887) (Anaphora) Brazil (Ce)
109 **misema** Walsingham, 1914 Mexico (Gue)
110 **modestus** Busck, 1912 Mexico (Pue)
111 **monoctenis** Meyrick, 1931 Paraguay
112 **morbidula** Meyrick, 1931 Brazil (SP)
113 **niveipunctata** Walsingham, 1892 Cuba
114 **noctivaga** (Walsingham, 1897) (Pseudanaphora) Grenada
115 **noctuina** (Walsingham, 1892) (Anaphora) Cuba
116 **nubifer** Walsingham, 1914 Panama
117 **numidia** (Druce, 1901) (Anaphora) Mexico (Ver)
118 **occultum** (Walsingham, 1897) (Atopocera) Haiti
119 **ochracea** (Möschler, 1890) (Caenogenes) Puerto Rico
120 **orasiusalis** (Walker, 1859) (Palthis) [Belize]
orasialis Walsingham, 1914, missp.
121 **orizabae** (Dyar, 1907) (Anaphora) Mexico (Ver)
122 **ornata** (Walsingham, 1887) (Anaphora) Colombia?
123 **pachynta** Meyrick, 1913 Colombia
124 **pallidus** Möschler, 1881 Surinam
125 **panamae** Busck, 1914 Panama
126 **pannephela** Meyrick, 1931 Brazil (ES)
127 **particeps** Meyrick, 1913 Venezuela
128 **parvus** (Walsingham, 1897) (Hypoclopus) Dominica
129 **pauper** Walsingham, 1914 Mexico (Ver)
130 **penumbra** Walsingham, 1914 Panama
131 **perissarcha** Meyrick, 1919 Guyana
132 **perpetua** Meyrick, 1913 Paraguay
133 **perrensella** (Walsingham, 1887) (Caenogenes) Argentina
134 **perrensi** (Druce, 1901) (Anaphora) Brazil (Go)
135 **phaeomalla** Meyrick, 1913 Argentina
136 **pinnifera** Meyrick, 1931 Brazil (MG?)
137 **poeyi** Walsingham, 1892 St. Vincent
*138 **popeanellus** (Clemens, 1859) (Anaphora) USA (Tx)
agrotipennella (Grote, 1872) (Anaphora) USA (Al)
scardina (Zeller, 1873) (Anaphora) USA (Tx)
morrisoni (Walsingham, 1887) (Anaphora) USA (Fl)
confusellus J. B. Smith, 1891 USA (Ga)
139 **practica** Meyrick, 1913 Brazil (MG?)
140 **pretusalis** (Walker, 1858) (Palthis) [Belize]
granulatella (Walker, 1863) (Tirasia) Brazil (Am)
praetusalis Walsingham, 1914, missp.
cathecta Walsingham, 1914, **n. syn.** Mexico (Ver)
141 **prepodes** Walsingham, 1914 Mexico (Gue)
142 **pristinella** (Walker, 1863) (Tinea) Venezuela
143 **psammophila** Meyrick, 1931 Brazil (MG?)
144 **pseudohirsutus** Hasbrouck, 1964, repl. name USA (Ca)
hirsutus Busck, 1912, preocc. (not Walsingham, 1887) USA (Ca)
145 **pseudonoma** Meyrick, 1922 Brazil (Am)
146 **psoloessa** Meyrick, 1932 Brazil (SC)
147 **pumicea** Meyrick, 1913 Paraguay
148 **punctata** (Druce, 1901) (Anaphora) Costa Rica
149 **pusilla** (Zeller, 1877) (Anaphora) Colombia
150 **pygmaea** (Walsingham, 1887) (Felderia) Argentina
151 **pyramellus** (Barnes & McDunnough, 1913) (Eulepiste) USA (Nv)
152 **rastricornis** Meyrick, 1932 Virgin Is. (St. Thomas)
153 **reflexa** (Fabricius, 1793) (Bombyx) [Virgin Is.?]
154 **ridicula** Meyrick, 1913 Surinam
155 **rupestris** (Walsingham, 1914) (Apoclisis) Jamaica
156 **sachari** Busck, 1913 Guyana
157 **sagaritis** Meyrick, 1932 Bolivia
158 **salvini** (Druce, 1901) (Thysanosedes [sic]) Panama
arcei (Druce, 1901) (Anaphora), **n. syn.** Panama
caprimulgus Walsingham, 1914, **n. syn.** Panama
159 **sarista** Meyrick, 1913 Venezuela
160 **satyrisca** Meyrick, 1927 Colombia
161 **schistodes** Meyrick, 1913 Venezuela
162 **scopodes** Meyrick, 1913 Venezuela
163 **scotera** Walsingham, 1914 Mexico (Gue)
164 **scotina** (Walsingham, 1914) (Amydria) Mexico (Gue)
165 **scrupulata** (Meyrick, 1913) (Orothyntis) Colombia
166 **seminigera** Meyrick, 1913 Venezuela
167 **sepulcralis** (Meyrick, 1912) (Brachysymbola) Argentina
*168 **serratus** Hasbrouck, 1964 USA (Az)
169 **setiacma** Meyrick, 1923 Brazil (Am)
170 **signatus** Busck, 1920 Mexico (Chia)
171 **spathista** Meyrick, 1919 Colombia
172 **spinifera** Meyrick, 1913 Paraguay
173 **subfusca** Meyrick, 1913 Argentina
174 **superstes** Walsingham, 1914 Guatemala
175 **suspensilis** Meyrick, 1913 Brazil (SC)
176 **synapta** Durrant, 1914 Mexico (Tab)
177 **tetrancyla** Meyrick, 1913 Peru
*178 **texanellus** (Chambers, 1878) (Anaphora) USA (Tx)
hulstellus Beutenmüller, 1887 USA (Fl)
barnesii (Dyar, 1900) (Atopocera) USA (Tx)
179 **thaminodes** Meyrick, 1919 Guyana
180 **tholomicta** Meyrick, 1929 Colombia
181 **torta** (Meyrick, 1922) (Orothyntis) Peru
182 **tretus** Kaye, 1925 Trinidad
183 **triatomella** (Walsingham, 1877) (Anaphora) Virgin Is. (St. Thomas)
184 **tricausta** Meyrick, 1913 Argentina
185 **trichosoma** Durrant, 1914 Guatemala
186 **umbratipalpis** (Walsingham, 1892) (Eulepiste) Dom. Rep.
187 **uncigera** (Walsingham, 1897) (Anaphora) Colombia
188 **uncispinis** Walsingham, 1914 Costa Rica
189 **underwoodi** Druce, 1901 Costa Rica
*190 **vanduzeei** Hasbrouck, 1964 USA (Tx)
191 **variabilis** (Walsingham, 1887) (Ortholophus) USA (Az)
192 **vauriei** Hasbrouck, 1964 USA (Tx)
193 **vespertilio** Meyrick, 1931 Brazil (ES)
194 **victrix** Walsingham, 1914 Guatemala
195 **vigia** Beutelspacher, 1969 Mexico (Ver)
196 **vitellus** Poey, 1832 Cuba
197 **walsinghami** Möschler, 1890 Puerto Rico
triformellus Forbes, 1930, **n. syn.** Puerto Rico
triformalis Forbes, 1930, missp.
198 **whitelyi** (Druce, 1901) (Anaphora) Guyana
199 **xylinella** (Walker, 1863) (Eddara) Jamaica
200 **zanclophora** Meyrick, 1922 Peru

AMYDRIA Clemens, 1859
Casape Walker, 1864
Amadria Chambers, 1875, missp.
Amadrya Chambers, 1875, missp.
Amydrya Kearfott, 1903, missp.
Neomeristis Meyrick, 1919, **n. syn.**

201 **abscensella** (Walker, 1863) (Tinea), **n. comb.** Venezuela
202 **anceps** Walsingham, 1914 Mexico (Gue)
socialis (Beutelspacher, 1977) (Acrolophus), **n. syn.** Mexico (Jal)
*203 **apachella** Dietz, 1905 USA (Az)
204 **arizonella** Dietz, 1905 USA (Az)
205 **margoriella** Dietz, 1905 USA (Tx)
marjorieella Dietz, 1905, missp.
margorieella Dietz, 1905, missp.
marjoriella Busck, 1906, missp.
206 **meridionalis** Walsingham, 1914 Costa Rica
207 **muricolor** Walsingham, 1914 Mexico (Dur)
208 **obliquella** Dietz, 1905 USA (Ca)
209 **pauculella** (Walker, 1864) (Casape) Venezuela
210 **pogonites** Walsingham, 1914 Mexico (Gue)
211 **taracta** Walsingham, 1914 Guatemala

ANTIPOLISTES Forbes, 1933

212 **anthracella** Forbes, 1933 Puerto Rico
latebrivora (Meyrick, 1935) (Tinea), **n. syn.** Trinidad

ATTICONVIVA Busck, 1934

213 **melichrosta** (Meyrick, 1922) (Tiquadra), **n. comb.** Brazil (Pa)
eidmannella Busck, 1934, **n. syn.** Brazil (RJ)

AUGOLYCHNA Meyrick, 1922

214 **septemstrigella** (Chambers, 1878) (Tinea) USA (Tx)

AXIAGASTA Meyrick, 1930

215 **stactogramma** Meyrick, 1930 Brazil (Pa)

BARYMOCHTHA Meyrick, 1922

216 **entherastis** Meyrick, 1922 Brazil (Am)/Guyana

BASANASCA Meyrick, 1922

217 **parcens** Meyrick, 1922 Brazil (Am)

BRITHYCEROS Meyrick, 1932

218 **dichroanthes** Meyrick, 1932 Brazil (Am)/Peru

BYTHOCRATES Meyrick, 1919

219 **drosocycla** Meyrick, 1919 Guyana

CHOROPLECA Durrant, 1914
Cyane Chambers, 1873, preocc. (C. Felder, 1861 [Nymphalidae])
Diachalastis Meyrick, 1920

220 **acrodisca** Meyrick, 1917 Guyana
221 **boviceps** Walsingham, 1914 Mexico (Gue)
222 **catorthota** Meyrick, 1917 Guyana
223 **discatella** Walker, 1864 Brazil (Am)

224 **germana** Walsingham, 1914 Mexico (Gue)
225 **isodisca** Meyrick, 1917 Guyana
226 **murenula** Meyrick, 1924 Peru
227 **panscia** Meyrick, 1917 Ecuador
228 **poecilta** Walsingham, 1914 Mexico (Gue)
229 **rhombifera** Meyrick, 1917 Guyana
230 **sublimis** Meyrick, 1917 Colombia
231 **terpsichorella** (Busck, 1910) (Cyane)[8] USA (Hi)
tetraglossa (Meyrick, 1920) (Diachalastis) Fiji
232 **tripudians** Meyrick, 1924 Peru
233 **zygoterma** Meyrick, 1917 Colombia/Ecuador

CLEPTICODES Meyrick, 1927

234 **hexaleuca** Meyrick, 1932 Virgin Is. (St. Thomas)

CLINOGRAPTIS Meyrick, 1932

235 **ogmodes** Meyrick, 1932 Brazil (Am)

CNISMORECTIS Meyrick, 1936

236 **choritica** Meyrick, 1936 Bolivia

COLPOCRITA Meyrick, 1930

237 **diptila** Meyrick, 1930 Brazil (Pa)

COMODICA Meyrick, 1880

238 **ordinata** Walsingham, 1914 Mexico (Tab)

COMPSOCRITA Meyrick, 1922

239 **florens** Meyrick, 1922 Brazil (Pa)

CRANAODES Meyrick, 1919

240 **prostylias** Meyrick, 1927 Colombia
241 **stereopa** Meyrick, 1919 Colombia

CREPIDOCHARES Meyrick, 1922

242 **subtigrina** Meyrick, 1922 Brazil (Am)

CRYPHIOTECHNA Meyrick, 1932

243 **ochracma** Meyrick, 1932 Peru

DASMOPHORA Meyrick, 1919

244 **xerospila** Meyrick, 1919 French Guiana

DEMOBROTIS Meyrick, 1892

245 **haliplancta** Meyrick, 1927 Bermuda

DIATAGA Walsingham, 1914

246 **compsacma** Meyrick, 1919 Guyana
247 **leptosceles** Walsingham, 1914 Mexico (Ver)

DORATA Busck, 1904
Pterolonche Walsingham, 1889, preocc. (Zeller, 1847 [Pterolonchidae])
Dorota Kearfott, 1907, missp.

248 **nigritella** Busck, 1920 Mexico (Sin)
249 **sagittella** Busck, 1913 Mexico (Pue)

DRASTEA Walsingham, 1914

250 **mexica** Walsingham, 1914 Mexico (Ver)

DYOTOPASTA Busck, 1907
Pseudoxylesthia Walsingham, 1907
Dystopasta McDunnough, 1939, emend.

251 **yumaella** Kearfott, 1907 USA (Az)
angustella (Walsingham, 1907) (Pseudoxylesthia) USA (Az/Tx)

DYSOPTUS Walsingham, 1914

252 **probata** Walsingham, 1914 Guatemala
253 **tantalota** Meyrick, 1919 Guyana

ECTINOCAMPA Silvestri, 1944

254 **nasutitermina** Silvestri, 1944 Brazil (SP)

EPHEDROXENA Meyrick, 1919

255 **incisoria** Meyrick, 1919 Guyana

EPISYRTA Meyrick, 1929

256 **coniomicta** Meyrick, 1929 Colombia
257 **protonistis** Meyrick, 1930 Brazil (Pa)

ERECHTHIAS Meyrick, 1880
Ereunetis Meyrick, 1880
Decadarchis Meyrick, 1886
Nesoxena Meyrick, 1929
Amphisyncentris Meyrick, 1933
Caryolestis Meyrick, 1934
Triadogona Meyrick, 1937
Anemerarcha Meyrick, 1937
Erechtias Ghesquière, 1940, missp.
Tinexotaxa Gozmány, 1968
Neodecadarchis Zimmerman, 1978
Lepidobregma Zimmerman, 1978
Pantheus Zimmerman, 1978

258 **darwini** Robinson, 1983 Brazil (St. Paul's Rocks)
259 **minuscula** (Walsingham, 1897) (Ereunetis) Virgin Is. (St. Croix/St. Thomas)
260 **zebrina** (Butler, 1881) (Argyresthia) USA (Hi)
lanceolata (Walsingham, 1897) (Ereunetis) Virgin Is. (St. Thomas)
xenica (Meyrick, 1911) (Ereunetis) Seychelles
travestita (Gozmány, 1968) (Tinexotaxa) Sierra Leone

EXONCOTIS Meyrick, 1919

261 **gemistis** (Meyrick, 1909) (Amydria), **n. comb.** Bolivia
262 **resona** Meyrick, 1929 Colombia
263 **umbraticella** (Busck, 1914) (Amydria) Panama
increpans Meyrick, 1919, **n. syn.** French Guiana

HOMODOXUS Walsingham, 1914

264 **aristula** Walsingham, 1914 Guatemala

HOMOSETIA Clemens, 1863

265 **anaphrictis** Meyrick, 1919 Guyana
266 **ancyropis** Meyrick, 1919 Guyana
267 **cosmopa** Meyrick, 1919 Guyana
268 **iambica** Meyrick, 1919 Colombia
269 **scandalitis** Meyrick, 1919 Guyana
270 **tephropis** Meyrick, 1919 Ecuador

HOMOSTINEA Dietz, 1905
Homotinea Meyrick, 1932, emend.

271 **chersadacta** Meyrick, 1932 (Homotinea [sic]) Brazil (Am/Pa)

HORMANTRIS Meyrick, 1927

272 **astragalopa** Meyrick, 1927 Colombia

HYBROMA Clemens, 1862

273 **anticosma** Meyrick, 1919 Colombia
274 **crocorrhoa** Meyrick, 1919 Colombia
275 **dulica** Walsingham, 1914 Mexico (Tab)
276 **paedisca** Walsingham, 1914 Mexico (Gue)
277 **pegaea** Meyrick, 1919 Guyana
278 **zacharis** Meyrick, 1919 Guyana

INFURCITINEA Spuler, 1910
Atinea Amsel, 1954
Microtinea Amsel, 1954

279 **luteela** Forbes, 1931 Puerto Rico
280 **palpella** Forbes, 1931 Puerto Rico

ISOCORYPHA Dietz, 1905
Socorypha Busck, 1909, missp.

281 **limbata** Walsingham, 1914 Mexico (Tab)

LEPTOCHERSA Meyrick, 1919

282 **diarthra** Meyrick, 1919 Guyana

LEPYROTICA Meyrick, 1921

283 **brevistrigata** (Walsingham, 1897) (Tinea), **n. comb.** Virgin Is. (St. Croix/St. Thomas)
284 **delotoma** (Meyrick, 1919) (Tinea), **n. comb.** French Guiana
285 **diluticornis** (Walsingham, 1897) (Tinea), **n. comb.** Virgin Is. (St. Thomas)
scythropiella (Walsingham, 1897), (Tinea), **n. syn.** Virgin Is. (St. Thomas)
286 **fragilella** (Walsingham, 1897) (Tinea) Haiti
287 **reduplicata** (Walsingham, 1897) (Tinea) Virgin Is. (St. Thomas)
288 **scardamyctis** Meyrick, 1921 West Indies ("Leeward Is.")

LEUCOPHASMA Walsingham, 1897

289 **carmodiella** Busck, 1910 Tobago
290 **phantasmella** Walsingham, 1897 Grenada

LINDERA Blanchard, 1852
Palpula Blanchard, 1852, preocc. (Treitschke, 1833 [Oecophoridae])
Safra Walker, 1864, preocc. (Walker, 1863 [Pyralidae])
Chresotes Butler, 1881
Paraneura Dietz, 1905
Cervitinea Amsel, 1956

291 **baliopa** Meyrick, 1917 Brazil ("Rio Tapenaya")
292 **onychias** Meyrick, 1931 Brazil (ES)
293 **venezuelensis** (Amsel, 1956) (Cervitinea) Venezuela
294 **tessellatella** Blanchard, 1852 Chile
variegella (Blanchard, 1852) (Palpula) Chile
bogotatella (Walker, 1864) (Safra) Colombia
simulella (Dietz, 1905) (Paraneura) USA (Ca)
ehrhornella (Dietz, 1905) (Paraneura) USA (Ca)
cruciferella (Dietz, 1905) (Paraneura) USA (Ca)

LIPOMERINX Walsingham, 1914

295 **prismatica** Walsingham, 1914 Mexico (Ver)
erebantha (Meyrick, 1919) (Trithamnora), **n. syn.** Guyana

LITHOPSAESTIS Meyrick, 1932

296 **mixophanes** Meyrick, 1932 Brazil (SC)

MEA Busck, 1906
Progona Dietz, 1905, preocc. (Berg, 1882 [Arctiidae])

297 **incudella** Forbes, 1931 Puerto Rico
298 **yunquella** Forbes, 1931 Puerto Rico

MONOPIS Hübner, [1825]
Blabophanes Zeller, 1852
Hyalospila Herrich-Schäffer, 1853
Rhitia Walker, 1864
Eusynopa Lower, 1903

299 **impressipenella** (Bilimek, 1867) (Ornix) Mexico (Mor)
cacahuamilpensis (Herrera, 1892) (Ornix) Mexico (Mor)
300 **prometopias** (Gyen, 1913) (Amydria), **n. comb.** Chile

MYRMECOZELA Zeller, 1852
Psecadioides Butler, 1881
Promasia Chrètien, 1905
Protolophe Rebel, 1915

301 **corymbota** Meyrick, 1919 Peru
302 **renitens** Meyrick, 1922 Brazil (Am)

MYTHOPLASTIS Meyrick, 1919

303 **chalcochra** Meyrick, 1931 Brazil (SP)
304 **exanthes** Meyrick, 1919 French Guiana

NEMAPOGON Schrank, 1802
Diaphthirusa Hübner, [1825]

305 **granella** (Linnaeus, 1758) (Tinea) [Sweden?]
costotristrigella (Chambers, 1873) (Tinea) USA (Ky)
fuscomaculella (Chambers, 1873) (Tinea) USA (Ky)
marmorella (Chambers, 1875) (Tinea) Canada
costistrigella (Dietz, 1905) (Tinea), missp.
fascomaculella (Dietz, 1905) (Tinea), missp.
nigroatomella (Dietz, 1905) (Tinea) USA (NJ)

NIDITINEA Petersen, 1957
Tineidia Zagulajev, 1960

306 **fuscella** (Linnaeus, 1758) (Tinea) [Sweden?]
spretella ([Denis & Schiffermüller], 1775) (Tinea) [Austria]
fuscipunctella (Haworth, 1828) (Tinea) UK (England)
nubilipennella (Clemens, 1859) (Tinea) USA [Pa]
abligatella (Walker, 1863) (Tinea) South Africa
ignotella (Walker, 1864) (Tinea) Sierra Leone
frigidella (Packard, 1867) (Oecophora) Canada
griseella (Chambers, 1873) (Tinea) USA (Ky)
eurinella (Zagulajev, 1952) (Tinea) USSR
distans (Gozmány, 1959) (Tinea) Afghanistan
307 **praeumbrata** (Meyrick, 1919) (Tinea) Guyana
negreai Căpuse & Georgescu, 1977 Cuba

OENOE Chambers, 1874

308 **euphrantis** Meyrick, 1927 Bermuda
309 **minimella** Forbes 1930 Virgin Is. (St. John)
310 **pumiliella** (Walsingham, 1897) (Tinea), **n. comb.** Virgin Is. (St. Croix/ St. Thomas)
311 **synchorda** Meyrick, 1919 Guyana

OINOPHILA Stephens, 1848
Oenophila Zeller, 1853, missp.

312 **v-flavum** (Haworth, 1828) (Gracillaria) UK (England)

OPOGONA Zeller, 1853
Lozostoma Stainton, 1859
Conchyliospila Wallengren, 1861
Cachura Walker, 1864
Dendroneura Walsingham, 1892
Hieroxestis Meyrick, 1892
Exala Meyrick, 1912

313 **hemidryas** Meyrick, 1915 Guyana
314 **leptynta** Meyrick, 1915 Guyana
315 **lotoxantha** Meyrick, 1915 Guyana
316 **metanastes** Walsingham, 1914 Mexico (Tab)
317 **pelinoma** Meyrick, 1931 Peru
318 **praestans** (Walsingham, 1892) (Dendroneura) St. Vincent
319 **pyrographa** Walsingham, 1914 Guatemala
320 **rhynchacma** Meyrick, 1920 Brazil (Am)
321 **serta** Walsingham, 1914 Mexico (Gue)
322 **simplex** (Walsingham, 1897) (Dendroneura), **n. comb.** Dom. Rep.
323 **subcervinella** (Walker, 1863) (Tinea) Mauritius
324 **xerota** Walsingham, 1914 Panama

OPSODOCA Meyrick, 1919

325 **amentata** Meyrick, 1919 Guyana
326 **metrodoxa** Meyrick, 1919 Guyana

OTOCHARES Meyrick, 1919

327 **gypsopa** Meyrick, 1919 Guyana
328 **peronacma** Meyrick, 1919 Guyana

PACHYDYTA Meyrick, 1922
Pachdyta Meyrick, 1922, missp.

329 **clitozona** Meyrick, 1922 Brazil (Pa)

PALAEPHATUS Butler, 1883

330 **falsus** Butler, 1883 Chile

PANTHYTARCHA Meyrick, 1922

331 **astrocharis** Meyrick, 1922 Brazil (Pa)

PEDALIOTIS Meyrick, 1930

332 **dryograpta** Meyrick, 1930 Brazil (Pa)

PERILICMETIS Meyrick, 1932

333 **diplaca** Meyrick, 1932 Brazil (SC)

PHAEOSES Forbes, 1922

334 **sabinella** Forbes, 1922 USA (La)

PHEREOECA Hinton & Bradley, 1956

335 **allutella** (Rebel, 1892) (Tineola) Canary Is.
336 **dubitatrix** (Meyrick, 1932) (Tinea), **n. comb.** Virgin Is. (St. Thomas)
walsinghami (Busck, [1934]) (Tineola), **n. syn.** Virgin Is. (St. Thomas)
337 **uterella** (Walsingham, 1897) (Tineola) Brazil (Pa)
borboropis (Meyrick, 1919), misid. [in part] Guyana

POLYPSECTA Meyrick, 1930

338 **halmeuta** Meyrick, 1930 Brazil (Pa)

POMPOSTOLELLA Fletcher, 1940
Pompostola Meyrick, 1927, preocc. (Hübner, [1819] [Zygaenidae])

339 **aeneoalbida** (Walsingham, 1897) (Ereunetis), **n. comb.** Virgin Is. (St. Thomas)
340 **charipepla** (Meyrick, 1927) (Pompostola) Bermuda

PRAEACEDES Amsel, 1954
Titaenoses Hinton & Bradley, 1956

341 **atomosella** (Walker, 1863) (Tinea) Sierra Leone
seminolella (Beutenmüller, 1889) (Tinea), **n. syn.** USA (Fl)
thecophora (Walsingham, 1908) (Tinea), **n. syn.** Canary Is.
despecta (Meyrick, 1919) (Tinea), **n. syn.** Guyana
deluccae Amsel, 1954, **n. syn.** Malta

PROBOLOPTILA Meyrick, 1921

No.	Species	Distribution
342	**aeolella** (Walsingham, 1897) (Ereunetis), **n. comb.**	Virgin Is. (St. Thomas)
343	**frontella** (Walsingham, 1897) (Ereunetis)	Virgin Is. (St. Thomas)

PROTODARCIA Forbes, 1931

No.	Species	Distribution
344	**argyrophaea** Forbes, 1931	Puerto Rico
345	**bicolorella** Forbes, 1931	Puerto Rico
346	plumella (Walsingham, 1892) (Tinea)	St. Vincent
347	**tetraonella** (Walsingham, 1897) (Tinea), **n. comb.**	Virgin Is. (St. Croix/ St. Thomas)
348	**tischeriella** (Walsingham, 1897) (Tinea), **n. comb.**	Haiti/Virgin Is. (St. Thomas)

PTILOPSALTIS Meyrick, 1935

No.	Species	Distribution
348.1	**synchorista** Meyrick, 1935	Trinidad

SCARDIA Treitschke, 1830
Agarica Sodoffsky, 1837
Morophaga Herrich-Schäffer, 1853
Atabyria Snellen, 1884
Osphretica Meyrick, 1910
Microscardia Amsel, 1952

No.	Species	Distribution
349	**bimendella** (Zeller, 1863) (Tinea), **n. comb.**	Venezuela
350	**brasiliensis** (Zagulajev, 1966) (Morophaga), **n. comb.**	Brazil (RS)
351	**funeratella** (Zeller, 1863) (Tinea), **n. comb.**	Venezuela
352	**isthmiella** Busck, 1914	Panama
353	**iulina** (Walsingham, 1914) (Phycis)	Guatemala
354	**luctuosa** (Walsingham, 1914) (Phycis)	Costa Rica/Mexico (Ver)
355	**minimella** Busck, 1914	Panama
356	**montium** (Walsingham, 1914) (Phycis)	Panama

SETIARCHA Meyrick, 1932

No.	Species	Distribution
357	**aleuropis** Meyrick, 1932	Brazil (SC)

SETOMORPHA Zeller, 1852
Epilegis Dietz, 1905
Apotomia Dietz, 1905
Semiota Dietz, 1905
Trisyntopa Lower, 1918

No.	Species	Distribution
358	**cycladopa** (Meyrick, 1927) (Syrmologa), **n. comb.**	Peru
359	**rutella** Zeller, 1852	South Africa
	insectella auct. (not Fabricius, 1794)	
	rupicella Zeller, 1852	Cuba
	operosella Zeller, 1873	USA (Tx)
	inamoenella Zeller, 1873	USA (Tx)
	ruderella Zeller, 1873	USA (Tx)
	multimaculella (Chambers, 1878) (Gelechia)	USA (Tx)
	dryas (Butler, 1881) (Chrestotes)	USA (Hi)
	corticinella Snellen, 1884	Indonesia
	bogotatella Alpheraky, 1889 (not Walker, 1864)	Canary Is.
	discipunctella Rebel, 1894	Canary Is.
	majorella Dietz, 1905	USA (Ca)
	sigmoidella Dietz, 1905	USA (Co)
	cariosella (Dietz, 1905) (Epilegis)	USA (Ca)
	fractiliniella (Dietz, 1905) (Apotomia)	USA (Ca)
	transversestrigella (Dietz, 1905) (Semiota)	USA (Ca)
	margalaestriata Keuchenius, 1917	Indonesia
	euryspoda (Lower, 1918) (Trisyntopa)	Australia
	tineoides Dammerman, 1919	Indonesia
	nitella Voute, 1935, missp.	

SYNCRATERNIS Meyrick, 1922

No.	Species	Distribution
360	**anthestias** Meyrick, 1922	Brazil (Am)
361	**phaeospila** Meyrick, 1922	Brazil (Am)

SYRMOLOGA Meyrick, 1919

No.	Species	Distribution
362	**chersopa** Meyrick, 1919	Guyana
363	**leucoclistra** Meyrick, 1919	Colombia
364	**spermatias** Meyrick, 1919	Guyana
365	**thriophora** Meyrick, 1919	Guyana

TAENIODICTYS Forbes, 1933

No.	Species	Distribution
366	**sericella** Forbes, 1933	Puerto Rico

TETRAPALPUS Davis, 1972

No.	Species	Distribution
367	**trinidadensis** Davis, 1972	Trinidad

TINEA Linnaeus, 1758[9]
Autoses Hübner, [1825]
Acedes Hübner, [1825]
Edosa Walker, 1866
Chrysoryctis Meyrick, 1886
Dystinea Börner, 1925

No.	Species	Distribution
368	**absolutrix** Meyrick, 1930	Peru
369	**analytica** Meyrick, 1919	Guyana
370	**auromaculata** Walsingham, 1897	Virgin Is. (St. Thomas)
371	**bivirgella** R. Felder & Rogenhofer, 1875	Colombia
	dividuella Zeller, 1877, **n. syn.**	Colombia
372	**boliviana** Busck, 1911	Bolivia
373	**borboropis** Meyrick, 1919	Guyana
	borboropis Meyrick, 1922, redesc.	Guyana
	barbaropis Busck, [1934], missp.	
374	**caducella** Zeller, 1877	Panama?
375	**caerula** Meyrick, 1927	Peru
376	**catalytica** Meyrick, 1919	Colombia
377	**chloroceros** Meyrick, 1919	Ecuador
378	**conchylitis** Meyrick, 1919	Peru
379	**coracopis** Meyrick, 1919	Peru
380	**cretella** Walsingham, 1897	Haiti
381	**crocodeta** Meyrick, 1927	Peru
382	**culminicola** (Staudinger, 1894) (Tineola)	Bolivia
383	**cumulatella** Zeller, 1877	Colombia
384	**dissimilis** Walsingham, 1914	Guatemala
385	**enchytopa** Meyrick, 1931	Argentina
386	**erasella** Zeller, 1863	Venezuela
387	**eriochrysa** Meyrick, 1932	Brazil (Pa)
388	**extracta** Meyrick, 1919	Guyana
389	**familiaris** Zeller, 1877	Colombia
390	**flectella** Walker, 1863	Venezuela
391	**frontestrigata** Walsingham, 1897	Virgin Is. (St. Croix)
392	**gypsomicta** Meyrick, 1931	Chile
393	**holocapna** Meyrick, 1931	Argentina
394	**isodonta** Meyrick, 1931	Chile
395	**latipennella** Zeller, 1877	Colombia
396	**minutella** (Fabricius, 1794) (Alucita)	"West Indies"
397	**montezuma** Meyrick, 1927	Bolivia
398	**murariella** Staudinger, 1859	Spain
	bipunctella (Ragonot, 1874) (Tineola)	Spain
399	**nigriceps** Zeller, 1877	Colombia
400	**nigripalpis** Walsingham, 1914	Mexico (Gue)
401	**ostiaria** Meyrick, 1927	Virgin Is. (St. Thomas)
	divisa Walsingham, 1897, preocc. (not Wollaston, 1879)	Virgin Is. (St. Thomas)
402	**oxymora** Meyrick, 1919	Peru
403	**pallescentella** Stainton, 1851	UK (England)
	nigrifoldella Gregson, 1856	UK (England)
	galeatella Mabille, 1888[10]	"Cape Horn" [Argentina?]
	horosema Meyrick, 1931	Argentina
	stimulatrix Meyrick, 1931	Argentina
404	**pallidorsella** Zeller, 1877	Colombia
405	**perisepta** Meyrick, 1922	Ecuador
406	**phaeonephela** Meyrick, 1927	Peru
407	**platysaris** Meyrick, 1931	Argentina
408	**praestabilis** Meyrick, 1927	Peru
409	**prensoria** Meyrick, 1931	Chile
410	**scotocleptes** Meyrick, 1934	Costa Rica
411	**scrutatricella** Zeller, 1877	Colombia
412	**sequens** Meyrick, 1919	Guyana
413	**solenobiella** Walsingham, 1897	Virgin Is. (St. Croix/St. Thomas)
414	**subcuprea** Meyrick, 1932	Brazil (Am)/Peru
415	**symmeles** (Meyrick, 1919) (Lampronia)	Colombia
416	**translucens** Meyrick, 1917	Pakistan
	metonella Pierce & Metcalfe, 1934	UK (England)
	leonhardi Petersen, 1957	Yugoslavia
	margaritacea Gozmány, 1967	Zaire
	fortificata Gozmány, 1968	South Africa
417	**umbraticostella** Walsingham, 1897	St. Vincent/Virgin Is. (St. Croix/St. Thomas)
418	**xanthosomella** (Maassen, 1890) (Gelechia), **n. comb.**	Bolivia
419	**xenodes** Meyrick, 1909	Bolivia

TIQUADRA Walker, 1863
Oscella Walker, 1864
Manchana Walker, 1866
Ventia Walker, 1866
Acureuta Zeller, 1877

No.	Species	Distribution
420	**aeneonivella** (Walker, 1864) (Oscella)	Venezuela
	aenconivella (Walker, 1864) (Oscella), missp.	
421	**avitella** Walker, 1866	Colombia
	ruderella (R. Felder & Rogenhofer, 1875) (Scardia)	Brazil (Am)
	aspera (Zeller, 1877) (Acureuta)	Colombia
422	**butyranthes** Meyrick, 1931	Brazil
423	**circumdata** (Zeller, 1877) (Acureuta)	Colombia
424	**crocidura** Meyrick, 1922	Brazil (SC)
425	**drapetica** Meyrick, 1919	Brazil (Pa)
426	**exercitata** Meyrick, 1922	Brazil (SC)
427	**galactura** Meyrick, 1931	Brazil (SP)
428	**inscitella** Walker, 1863	"Mexico"
429	**lentiginosa** (Zeller, 1877) (Acureuta)	Brazil (RJ)/Peru
430	**mallodeta** Meyrick, 1924	Mexico (Chia)
431	**nivosa** (R. Felder & Rogenhofer, 1875) (Scardia)	Brazil (RJ)
432	**nubilella** Amsel, 1956	Venezuela
433	**nucifraga** Meyrick, 1919	Colombia

434 **pirchuniae** (Zeller, 1877) (Acureuta) Argentina
435 **pontifica** Meyrick, 1919 French Guiana
436 **reversella** (Walker, 1866) (Ventia) Brazil (Am)
437 **semiglobata** Meyrick, 1922 Peru
438 **stenopa** Walsingham, 1914 Guatemala
439 **syntripta** Meyrick, 1922 Brazil (Pa)
440 **vilis** Meyrick, 1922 Brazil (Am)
coracophila Meyrick, 1932 Argentina

TRICHOPHAGA Ragonot, 1894

441 **tapetzella** (Linnaeus, 1758) (Tinea) [Sweden?]
tapetiella (Zeller, 1852) (Tinea), missp.
palaestrica (Butler, 1877) (Tinea) New Zealand

TRIEROSTOLA Meyrick, 1932

442 **remivola** Meyrick, 1932 Brazil (Am)

XYLESTHIA Clemens, 1859
Xylestia Dyar, [1903], missp.

443 **horridula** (Zeller, 1877) (Ochsenheimeria) Colombia
444 **menidias** Meyrick, 1922 Brazil (Am/Pa)/Peru
445 **pruniramiella** Clemens, 1859 USA [Pa]
clemensella Chambers, 1873 USA [Ky]
congeminatella Zeller, 1873 USA (Ma)
kearfottella Dietz, 1905 USA (NJ/Pa)

XYSTROLOGA Meyrick, 1919
Achanodes Meyrick, 1922, **n. syn.**
Syrrhoaula Meyrick, 1932, **n. syn.**

446 **antipathetica** (Forbes, 1931) (Achanodes), **n. comb.** Puerto Rico
447 **fulvicolor** Meyrick, 1919 Guyana
448 **grenadella** (Walsingham, 1897) (Setomorpha), **n. comb.** Grenada
449 **invidiosa** Meyrick, 1919 Colombia
450 **lactirivis** (Meyrick, 1932) (Syrrhoaula), **n. comb.** Brazil (Am)
451 **nigrovittata** (Walsingham, 1897) (Tinea), **n. comb.** Virgin Is. (St. Thomas)
452 **sympathetica** (Meyrick, 1922) (Achanodes), **n. comb.** Brazil (Am)

ZONOCHARES Meyrick, 1922

453 **tetradyas** Meyrick, 1922 Brazil (Am)

ZYMOLOGA Meyrick, 1919

454 **mylicopa** Meyrick, 1919 Colombia

16. PSYCHIDAE

by D. R. Davis

Penestoglossinae

AMIANTASTIS Meyrick, 1932
Autocnaptis Meyrick, 1935

1 **manicola** Meyrick, 1932 Argentina
2 **brachycasis** Meyrick, 1932 Argentina
sciospora (Meyrick, 1935) (Autocnaptis) Argentina
3 **renovata** (Meyrick, 1922) (Narycia), **n. comb.** Brazil (Am)

PLUMANA Busck, 1911
Homilostola Meyrick, 1917

4 **aequanima** (Meyrick, 1917) (Homilostola) Peru
5 **ascalopa** (Meyrick, 1917) (Homilostola) French Guiana
6 **autoplecta** (Meyrick, 1917) (Homilostola) French Guiana
7 **piperatella** Busck, 1911 French Guiana
8 **taeniata** (Meyrick, 1917) (Homilostola) French Guiana

PERISCEPTIS Meyrick, 1931

9 **horiarcha** Meyrick, 1931 Paraguay

PTEROGYNE Davis, 1975

10 **insularis** Davis, 1975 Dom. Rep.

Psychinae

EPICHNOPTERIX Hübner, [1825]
Epichnopteryx Heylaerts, 1875, missp.

11 **pulla** (Esper, 1785) (Bombyx) [Europe]
fiebrigi (Köhler, 1939) (Cochliotheca) Paraguay

PAUCIVENA Davis, 1975

12 **reticulata** Davis, 1975 Puerto Rico
13 **hispaniolae** Davis, 1975 Dom. Rep.

PROCHALIA Barnes & McDunnough, 1913

14 **licheniphilus** (Köhler, 1939) (Chlania [sic]), **n. comb.** Cuba

METAXYPSYCHE Davis, 1975

15 **trinidadensis** Davis, 1975 Trinidad

NAEVIPENNA Davis, 1964

16 **aphaidropa** (Dyar, 1914) (Platoeceticus) Panama
17 **cruttwellae** Davis, 1975 Trinidad
cruttwelli Davis, 1975, missp.

CRYPTOTHELEA Duncan, 1841
Platoeceticus Packard, 1869
Cryptotheles Costa Lima, 1945, missp.

18 **surinamensis** (Möschler, 1878) (Psyche) Surinam
19 **watsoni** (Jones, 1923) (Psyche) Haiti
20 **macleayi** (Guilding, 1827) (Oiketicus) "West Indies" [St. Vincent]
macleaii (Westwood, 1854) (Oiketicus), missp.
macleayii (Meyrick & Lower, 1907) (Eurycyttara [sic]), missp.
21 **hoffmanni** (Köhler, 1939) (Platoeceticus) Cuba
22 **congregata** (Jones, 1945) (Platoeceticus) Mexico (Mor)
23 **symmicta** (Dyar, 1914) (Platoeceticus) Panama
24 **gloverii** (Packard, 1869) (Platoeceticus) USA (Fl)
jonesi (Barnes & Benjamin, 1922) (Maratha) USA (Tx)
pizote (Schaus, 1927) (Chalia) Guatemala

DENDROPSYCHE Jones, 1926

25 **venezuelae** Davis, 1975 Venezuela
26 **burrowsi** Jones, 1926 Guyana

LUMACRA Davis, 1964

27 **brasiliensis** (Heylaerts, 1884) (Eumeta) Brazil
marona (Schaus, 1905) (Platoeceticus) French Guiana
costaricensis (Schaus, 1911) (Platoeceticus) Costa Rica
28 **kuenckelii** (Heylaerts, 1901) (Chalia) Argentina
künckelii (Heylaerts, 1901) (Chalia), incorr. spell.
küenkeli (Köhler, 1924) (Chalia), missp.
künckeli (Köhler, 1939) (Chalia), missp.
29 **haitiensis** Davis, 1964 Haiti
30 **quadridentata** Davis, 1964 French Guiana
31 **hyalinacra** Davis, 1964 El Salvador
32 **leucobasilaris** Davis, 1975 Venezuela

CURTORAMA Davis, 1964

33 **cassiae** (Weyenbergh, 1884) (Platoeceticus) Argentina
rebeli (Köhler, 1924) (Chalia) Argentina
rugosus (Köhler, 1931) (Platoeceticus) Argentina

ASTALA Davis, 1964

34 **confederata** (Grote & C. Robinson, 1868) (Psyche) USA (Tx)
lepidopteris (Dyar, 1926) (Pachythelia) Mexico (Col)
35 **hoffmanni** (Vázquez, 1941) (Eurukuttarus) Mexico (Pue)
36 **polingi** (Barnes & Benjamin, 1924) (Eurukuttarus) USA (Az)
37 **tristis** (Schaus, 1901) (Chalia) Mexico (Ver)
38 **zacualpania** (Dyar, 1916) (Chalia) Mexico (Mor?)
39 **vigasi** (Schaus, 1901) (Chalia) Mexico (Ver)

Oiketicinae

THANATOPSYCHE Butler, 1882

40 **chilensis** (Philippi, 1860) (Psyche) Chile
41 **canescens** Butler, 1882 Chile

ANIMULA Herrich-Schäffer, 1856
Subgenus **Animula** Herrich-Schäffer, 1856

42 **limpia** Dognin, 1894 Ecuador
43 **microptera** (Schaus, 1905) (Thyridopteryx) French Guiana
44 **dichroa** Herrich-Schäffer, 1856 Venezuela

Subgenus **Artipenna** Davis, 1964

45 **seitzi** (Gaede, 1936) (Thyridopteryx) Brazil (RJ)

BIOPSYCHE Dyar, 1905

46 **thoracica** (Grote, 1865) (Hymenopsyche) Cuba
47 **apicalis** (Hampson, 1904) (Thanatopsyche) Bahamas

OIKETICUS Guilding, 1827
Oeketicus Lefebre, 1842, missp.
Oeceticus Herrich-Schäffer, 1858, missp.
Oiketikus Seitz, 1919, missp.
Oicocestis Köhler, 1924, missp.
Oiceticus Lahille, 1926, missp.

Subgenus **Paraoiketicus** Davis, 1964

48 **bergii** (Weyenbergh, 1884) (Psyche) Argentina
oviformis Köhler, 1939 Argentina
49 **borsanii** Köhler, 1953 Argentina
50 **geyeri** Berg, 1877 "Patagonia" [Argentina]
tabacillus Weyenbergh, 1884 Argentina
jonesi Schaus, 1896 Brazil (SP)
thoracica (Schaus, 1905) (Thanatopsyche) Guyana
51 **zihuatanejensis** Vázquez, 1951 Mexico (Gue)

Subgenus **Oiketicus** Guilding, 1827

52 **toumeyi** Jones, 1922 USA (Az)
mortonjonesi Vázquez, 1949 Mexico (Mex)
53 **assimilis** Vázquez, 1942 Mexico (Mic?/Mor?)
54 **specter** Schaus, 1905 Venezuela
55 **townsendi** Townsend, 1894
* a) **townsendi** Townsend, 1894 USA (NM)
bonniwelli Barnes & Benjamin, 1924 USA (Tx)
* b) **dendrokomos** Jones, 1926 USA (Tx)
c) **mexicanus** Gaede, 1936 Mexico
multidentatus Vázquez, 1942 Mexico (Ver)
*56 **abbotii** Grote, 1880 USA (Tx)
abboti Holland, 1905, missp.
57 **kirbyi** Guilding, 1827 "West Indies" [St. Vincent]
kirbii Walker, 1855, missp.
fulgurator Herrich-Schäffer, 1856 Brazil
fulgerator Grote, 1865 (Oeceticus [sic]), missp.
poeyi Lucas, 1857 Cuba
gigantea (Zeller, 1871) (Psyche) Brazil (Pe)
orizavae Schaus, 1901 Mexico (Ver)
sinaloanus Vázquez, 1942 Mexico (Sin)
ochoterenai Vázquez, 1942 Mexico (Sin)
fasciculatus Vázquez, 1942 Mexico (Ver)
58 **platensis** Berg, 1883 Argentina
kerbyi Lahille, 1926 (Oiceticus [sic]), missp.
(not Guilding, 1827)

THYRIDOPTERYX Stephens, 1834
Hymenopsyche Grote, 1865

*59 **meadi** H. Edwards, 1881 USA (Ca)
meadii Kirby, 1892, missp.
60 **ephemeraeformis** (Haworth, 1803) (Sphinx) UK [England]
coniferarum (Packard, 1864) (Oeceticus [sic]) USA
pallidovenata Grossbeck, 1917 USA (Fl)
vernalis Jones, 1923 USA (De)

Unplaced Species

61 "Psychoglene" **basinigra** R. Felder, 1874 Brazil
62 "Psyche" **burmeisteri** Weyenbergh, 1884 Argentina
63 "Platoeceticus" **chaquensis** Köhler, 1939 Argentina
64 "Chalia" **daguerrei** Köhler, 1939 Argentina
65 "Chalia" **dispar** Köhler, 1939 Argentina
66 "Oiketicus" **elegans** Köhler, 1931 Argentina
67 "Oiketicus" **ginocchionus** Köhler, 1952 Argentina
68 "Zamopsyche" **haywardi** Köhler, 1939 Argentina
69 "Oiketicus" **horni** Köhler, 1938 Argentina
70 "Oiketicus" **lizeri** Köhler, 1939 Argentina
71 "Platoeceticus" **tandilensis** Köhler, 1931 Argentina
72 "Oiketicus" **westwoodii** Berg, 1882 Argentina
73 "Clania" **yamorkinei** Köhler, 1953 Argentina

17. ARRHENOPHANIDAE

by D. R. Davis

ARRHENOPHANES Walsingham, 1913

1 **perspicilla** (Stoll, 1790) (Bombyx) Surinam
inca Meyrick, 1913 Peru
2 **volcanica** Walsingham, 1913 Panama

CNISSOSTAGES Zeller, 1863

3 **mastictor** Bradley, 1951 Costa Rica
4 **oleagina** Zeller, 1863 Venezuela

ECPATHOPHANES Bradley, 1951

5 **anachoreta** Bradley, 1951 Colombia
6 **chiquita** (Busck, 1914) (Arrhenophanes) Panama

HARMACLONA Busck, 1914 [11]
Ptychoxena Meyrick, 1916

7 **cossidella** Busck, 1914 Panama

18. AMPHITHERIDAE

by J. B. Heppner

DASYCAREA Zeller, 1877 [12]

1 **viridisquamata** Zeller, 1877 Colombia

19. LYONETIIDAE

by D. R. Davis & S. E. Miller

Bedelliinae

EUPRORA Busck, 1906

1 **argentiliniella** Busck, 1906 USA (Tx)

PHILONOME Chambers, 1874
Phillonome Chambers, 1875, missp.
Eurynome Chambers, 1875, preocc. (Leach, 1814 [Crustacea])
Busckia Dyar, [1903]

2 **cuprescens** Walsingham, 1914 Mexico (Gue)
3 **spectata** Meyrick, 1920 Brazil (Am)

BEDELLIA Stainton, 1849

4 **somnulentella** (Zeller, 1847) (Lyonetia) Italy

Lyonetiinae

LYONETIA Hübner, [1825]
Argyromiges Curtis, 1829
Argyromis Stephens, 1829
Lyoneta Matsumura, 1931, missp.

5 **myura** Meyrick, 1931 Peru

Cemiostominae

LEUCOPTERA Hübner, [1825]
Cemiostoma Zeller, 1848

6 **entemopa** Walsingham, 1914 Mexico (Gue)
7 **salicis** Walsingham, 1914 Mexico (DF)

PERILEUCOPTERA Silvestri, 1943

8 **coffeella** (Guérin-Méneville, 1842) (Elachista) Guadeloupe?
noctuella (Madinier, 1870), unavail. (Art.11[13]) Martinique

COMPSOSCHEMA Walsingham, 1897

9 **bimarginellum** Walsingham, 1897 Virgin Is. (St. Thomas)

HAPALOTHYMA Meyrick, 1920

10 **ioplocama** Meyrick, 1920 Guyana
11 **xanthochorda** Meyrick, 1920 Guyana

THOMICTIS Meyrick, 1920

12 **ephorista** Meyrick, 1920 Guyana

Bucculatriginae

BUCCULATRIX Zeller, 1848
Ceroclastis Zeller, 1848
Bacculatrix Flint, Noble, & Shaw, 1978, missp.

13 **flexuosa** Walsingham, 1897 Virgin Is. (St. Thomas)
14 **gossypiella** Morrill, 1927 Mexico (Son)
15 **rhombophora** Meyrick, 1926 Bermuda
16 **stictopus** Walsingham, 1914 Mexico (Tab)
17 **subnitens** Walsingham, 1914 Mexico (Tab)
18 **thurberiella** Busck, 1914 USA (Az)
19 **unipuncta** Walsingham, 1897 Virgin Is. (St. Thomas)

20. GRACILLARIIDAE

by D. R. Davis & S. E. Miller

Gracillariinae

CALOPTILIA Hübner, [1825]
Poeciloptilia Hübner, [1825]
Ornix Treitschke, 1833
Coriscium Zeller, 1839
Antiolopha Meyrick, 1894

1 **aeneocapitella** (Walsingham, 1891) (Gracilaria [sic]), **n. comb.** St. Vincent
2 **aeolastis** (Meyrick, 1920) (Gracilaria [sic]), **n. comb.** Brazil (Am)
3 **burserella** (Busck, 1900) (Gracilaria [sic]) USA (Fl)
4 **callichora** (Meyrick, 1915) (Gracilaria [sic]), **n. comb.** Guyana
5 **camaronae** (Zeller, 1877) (Gracilaria [sic]), **n. comb.** Colombia

6 **chloroptilia** (Meyrick, 1915) (Gracilaria [sic]), **n. comb.** Guyana
7 **eolampis** (Meyrick, 1915) (Gracilaria [sic]), **n. comb.** Guyana
8 **hexameris** (Meyrick, 1921) (Gracilaria [sic]), **n. comb.** Brazil (Am)
9 **immuricata** (Meyrick, 1915) (Gracilaria [sic]), **n. comb.** Peru
10 **oriarcha** (Meyrick, 1915) (Gracilaria [sic]), **n. comb.** Peru
11 **pastranai** (Bourquin, 1962) (Gracilaria [sic]), **n. comb.** Argentina
12 **perseae** (Busck, 1920) (Gracilaria [sic]) USA (Fl)
13 **phiaropis** (Meyrick, 1921) (Gracilaria [sic]), **n. comb.** Peru
14 **pneumatica** (Meyrick, 1920) (Gracilaria [sic]), **n. comb.** Brazil (Pa)
15 **semiclausa** (Meyrick, 1921) (Gracilaria [sic]), **n. comb.** Brazil (Am)
16 **similatella** (Zeller, 1877) (Gracilaria [sic]), **n. comb.** Colombia
17 **viridula** (Zeller, 1877) (Gracilaria [sic]), **n. comb.** Colombia

EUCOSMOPHORA Walsingham, 1897

18 **cupreella** Walsingham, 1897 Jamaica
19 **dives** Walsingham, 1897 Grenada
20 **ornata** Walsingham, 1897 Grenada

NEUROSTROTA Ely, 1918
Neurostrata Ely, 1918, missp.

21 **gunniella** (Busck, 1906) (Acrocercops) USA (Tx)
22 **pithecolobiella** Busck, [1934] (Neurostrata [sic]) Cuba

PARECTOPA Clemens, 1860
Euspilopteryx Zeller, 1847, emend. (not Stephens, 1835)
Euspilapteryx Spuler, 1910, missp. (not Stephens, 1835)

23 **dactylota** Meyrick, 1915 Ecuador
24 **exorycha** Meyrick, 1928 Brazil (RS)
25 **heptametra** Meyrick, 1915 Colombia
26 **lithocolletina** (Zeller, 1877) (Gracilaria [sic]) Colombia
27 **lithomacha** Meyrick, 1915 Ecuador
28 **nesitis** (Walsingham, 1897) (Gracilaria [sic]) Virgin Is. (St. Thomas)
29 **pselaphotis** Meyrick, 1915 Ecuador
30 **pulverella** (Walsingham, 1897) (Gracilaria [sic]) Dom. Rep./Virgin Is.
31 **quadristrigella** (Zeller, 1877) (Gracilaria [sic]) Colombia
32 **refulgens** Meyrick, 1915 Ecuador
33 **rotigera** Meyrick, 1931 Chile
34 **trichophysa** Meyrick, 1915 Peru
35 **undosa** (Walsingham, 1897) (Gracilaria [sic]) Haiti/Virgin Is.
36 **viminea** Meyrick, 1915 Peru

NEUROBATHRA Ely, 1918

37 **curcassi** Busck, [1934] Cuba

PENICA Walsingham, 1914

38 **peritheta** Walsingham, 1914 Mexico (Gue)

PARORNIX Spuler, 1910

39 **errantella** (Walsingham, 1897) (Ornix) Virgin Is. (St. Thomas)
40 **micrura** Walsingham, 1914 Mexico (Gue)

CHILOCAMPYLA Busck, 1900

41 **psidiella** Busck, [1934] Cuba

DIALECTICA Walsingham, 1897
Eutrichocnemis Spuler, 1910

42 **permixtella** Walsingham, 1897 Dom. Rep./Grenada
43 **rendalli** Walsingham, 1897 Jamaica
44 **sanctaecrucis** Walsingham, 1897 Virgin Is.

ACROCERCOPS Wallengren, 1881

45 **achnodes** Meyrick, 1915 Ecuador
46 **albomarginata** (Walsingham, 1897) (Coriscium) Virgin Is. (St. Thomas)
47 **anthogramma** Meyrick, 1921 Brazil (Am)
48 **apicepunctella** (Walsingham, 1891) (Gracilaria [sic]) St. Vincent
49 **argocosma** Meyrick, 1915 Ecuador
50 **asaphogramma** Meyrick, 1920 Brazil (Pa)
51 **atalantis** Meyrick, 1924 Costa Rica
52 **attenuata** (Walsingham, 1897) (Coriscium) Virgin Is. (St. Thomas)
53 **breyeri** Bourquin, 1962 Argentina
54 **caementosa** Meyrick, 1915 Peru
55 **camptochrysa** Meyrick, 1921 Brazil (Am)
56 **chalinopa** Meyrick, 1920 Brazil (Pa)
57 **charitopis** Meyrick, 1915 Guyana
58 **chloronympha** Meyrick, 1921 Brazil (Am)
59 **chrysocosma** Meyrick, 1915 Guyana
60 **cirrhantha** Meyrick, 1915 Guyana
61 **cissiella** Busck, [1934] Cuba
62 **clitoriella** Busck, [1934] Cuba
63 **clytosema** Meyrick, 1920 Brazil (Am)
64 **contorta** Meyrick, 1920 Brazil (Pa)
65 **cordiella** Busck, [1934] Cuba
66 **crotalistis** Meyrick, 1915 Peru
67 **cyclogramma** Meyrick, 1921 Peru
68 **cymella** Forbes, 1931 Puerto Rico
69 **demotes** Walsingham, 1914 Mexico (Tab)
70 **desmochares** Meyrick, 1921 Brazil (Am)
71 **eclampsis** Durrant, 1914 Panama
72 **encentris** Meyrick, 1915 Guyana
73 **eurychalca** Meyrick, 1920 Brazil (Pa)
74 **fasciculata** Meyrick, 1915 Guyana
75 **gemmans** Walsingham, 1914 Mexico (Gue)
76 **hapsidota** Meyrick, 1915 Guyana
77 **hastigera** Meyrick, 1915 Ecuador
78 **helicomitra** Meyrick, 1924 Brazil (Ba)
79 **hippuris** Meyrick, 1915 Peru
80 **inconspicua** Forbes, 1930 Puerto Rico
81 **insulella** (Walsingham, 1892) (Zarathra) St. Vincent
82 **ipomoeae** Busck, [1934] Cuba
83 **leucographa** Clarke, 1953 Argentina
84 **leuconota** (Zeller, 1877) (Gracilaria [sic]) Colombia
85 **luctuosa** Meyrick, 1915 Guyana
86 **maranthaceae** Busck, [1934] Cuba
87 **marmaritis** Walsingham, 1914 Mexico (Gue)
88 **melanactis** Meyrick, 1915 Guyana
89 **melanocosma** Meyrick, 1920 Brazil (Pa)
90 **melantherella** Busck, [1934] Cuba
91 **microphis** Meyrick, 1921 Brazil (Am)
92 **nolickeniella** (Zeller, 1877) (Gracilaria [sic]) Colombia
93 **obversa** Meyrick, 1915 Guyana
94 **perturbata** Meyrick, 1921 Peru/Brazil (Am)
95 **piligera** Meyrick, 1915 Colombia
96 **pontifica** Forbes, 1931 Puerto Rico
97 **pylonias** Meyrick, 1921 Peru
98 **ramigera** Meyrick, 1920 Brazil (Pa)
99 **rhynchograpta** Meyrick, 1920 Brazil (Pa)
100 **serrigera** Meyrick, 1915 Ecuador/Peru
101 **sortis** Meyrick, 1915 Ecuador
102 **stalagmitis** Meyrick, 1915 Guyana
103 **taeniarcha** Meyrick, 1932 Brazil (Am)
104 **trimetala** Meyrick, 1915 Guyana
105 **undifraga** Meyrick, 1931 Haiti
106 **urbanella** (Zeller, 1877) (Gracilaria [sic]) Colombia
107 **xeniella** (Zeller, 1877) (Gracilaria [sic]) Colombia
108 **xystrota** Meyrick, 1915 Guyana
109 **zebrulella** Forbes, 1931 Puerto Rico

CUPHODES Meyrick, 1897
Phrixosceles Meyrick, 1908

110 **paragrapta** (Meyrick, 1915) (Phrixosceles), **n. comb.** Guyana

SPANIOPTILA Walsingham, 1897

111 **cordicaria** Meyrick, 1920 Brazil (Pa)
112 **eucnemis** Walsingham, 1914 Mexico (Gue)
113 **nemeseta** Meyrick, 1920 Brazil (Am)
114 **spinosum** Walsingham, 1897 Virgin Is. (St. Thomas)

LEUCANTHIZA Clemens, 1859

115 **forbesi** Bourquin, 1962 Argentina

MARMARA Clemens, 1863
Aesyle Chambers, 1875

116 **affirmata** (Meyrick, 1918) (Parectopa) Peru
117 **ischnotoma** (Meyrick, 1915) (Parectopa) Guyana
118 **isortha** (Meyrick, 1915) (Parectopa) Guyana
119 **phaneropis** (Meyrick, 1915) (Parectopa) Ecuador
120 **stemonodes** (Meyrick, 1915) (Parectopa) Ecuador

Lithocolletinae

CREMASTOBOMBYCIA Braun, 1908

121 **lantanella** Busck, 1910 [14] USA (Hi)

PHYLLONORYCTER Hübner, [1822]
Lithocolletis Hübner, [1825]
Eucestis Hübner, [1825]
Lithcolletes Matsumura, 1907, missp.
Phyllorycter Walsingham, 1914, missp.

122 **albimacula** (Walsingham, 1897) (Lithocolletis), **n. comb.** Virgin Is. (St. Thomas)
123 **antitoxa** (Meyrick, 1915) (Lithocolletis), **n. comb.** Peru
124 **argentifrontella** (Walsingham, 1897) (Lithocolletis) Virgin Is. (St. Thomas)
125 **chalcobaphes** Walsingham, 1914 Mexico (Gue)
126 **clerotoma** (Meyrick, 1915) (Lithocolletis), **n. comb.** Ecuador
126.1 **durangensis** Deschka, 1982 Mexico (Dur)
127 **epispila** (Meyrick, 1915) (Lithocolletis), **n. comb.** Ecuador
128 **iriphanes** (Meyrick, 1915) (Lithocolletis), **n. comb.** Peru
129 **oxygrapta** (Meyrick, 1915) (Lithocolletis), **n. comb.** Peru
130 **pictus** Walsingham, 1914 Mexico (Gue)
131 **solani** (E. M. Hering, 1958) (Lithocolletis), **n. comb.** Argentina
132 **stigmaphyllae** Busck, [1934] Cuba

133	**tenuicaudella** (Walsingham, 1897) (Lithocolletis), **n. comb.**	Virgin Is. (St. Croix)

CAMERARIA Chapman, 1902

*134	**agrifoliella** (Braun, 1908) (Lithocolletis)	USA (Ca)

PORPHYROSELA Braun, 1908

135	**desmodiella** (Clemens, 1859) (Lithocolletis)	USA (Pa)
	gregariella (Murtfeldt, 1881) (Lithocolletis)	USA (Mo)
136	**minuta** Clarke, 1953	Argentina

Phyllocnistinae

PHYLLOCNISTIS Zeller, 1848
Phylloenistis Chambers, 1875, missp.
Phylloetis Chambers, 1876, missp.
Phyllocnitis Busck, 1900, missp.

137	**abatiae** E. M. Hering, 1958	Argentina
138	**auriinea** Zeller, 1877	Colombia
139	**baccharidis** E. M. Hering, 1958	Argentina
140	**bourquini** Pastrana, 1960	Argentina
141	**dorcas** Meyrick, 1915	Guyana
142	**meliacella** Becker, 1974	Costa Rica
143	**rotans** Meyrick, 1915	Ecuador
[144	**rubella** (Blanchard, 1852) (Elachista) [15]	Chile]
145	**sciophanta** Meyrick, 1915	Peru
146	**sexangula** Meyrick, 1915	Peru
147	**wygodzinskyi** E. M. Hering, 1958	Argentina

Gelechioidea

21. OECOPHORIDAE

by V. O. Becker [16]

Depressariinae

Tribe Depressariini

EXAERETIA Stainton, 1849[17]
Depressariodes Turati, 1924
Martyrhilda Clarke, 1941

1	**ammitis** (Meyrick, 1931) (Depressaria)	Argentina
2	**ascetica** (Meyrick, 1926) (Depressaria)	Colombia
3	**baleni** (Zeller, 1877) (Depressaria)	Colombia
4	**lusciosa** (Meyrick, 1915) (Depressaria)	Peru
5	**mesosceptra** (Meyrick, 1915) (Depressaria)	Peru
6	**relegata** (Meyrick, 1922) (Depressaria)	Chile (Juan Fernandez Is.)
7	**rubristricta** (Walsingham, 1912) (Agonopteryx [sic])	Guatemala
8	**significa** (Meyrick, 1915) (Depressaria)	Ecuador

TALITHA Clarke, 1978 [18]

9	**anomala** Clarke, 1978	Chile

Tribe Amphisbatini

AFDERA Clarke, 1978

10	**orphnaea** (Meyrick, 1931) (Cryptolechia)	Chile

ANCIPITA Busck, 1914

11	**atteria** Busck, 1914	Panama

ATHRINACIA Walsingham, 1911

12	**cosmophragma** Meyrick, 1922	Brazil (Pa)
13	**leucographa** Walsingham, 1911	Mexico (Gue)
14	**psephophragma** Meyrick, 1929	Brazil (Am)
15	**trifasciata** Walsingham, 1911	Mexico (Gue)
16	**xanthographa** Walsingham, 1911	Mexico (Gue)

AUXOTRICHA Meyrick, 1931

17	**ochrogypsa** Meyrick, 1931	Peru

CHARIPHYLLA Meyrick, 1921

18	**closterias** Meyrick, 1921	Peru

COMOTECHNA Meyrick, 1920

19	**corculata** Meyrick, 1921	Brazil (Pa)
20	**dentifera** Meyrick, 1921	Peru
21	**ludicra** Meyrick, 1920	Guyana
22	**parmifera** Meyrick, 1921	Brazil (Pa)
23	**scutulata** Meyrick, 1921	Brazil (Am)
24	**semiberbis** Meyrick, 1921	Peru

COMPSISTIS Meyrick, 1888

25	**caerulipalpis** Meyrick, 1921	Peru
26	**homochorda** Meyrick, 1921	Peru
27	**labyrinthias** Meyrick, 1921	Peru
28	**macrochorda** Meyrick, 1921	Peru
29	**malacoscia** Meyrick, 1921	Peru

COPTOTELIA Zeller, 1863
Hyphypena Warren, 1889

30	**allardi** Clarke, 1951	Peru
31	**bipunctalis** (Warren, 1889) (Hyphypena)	Brazil (Am)
	byrsocyma (Meyrick, 1921) (Hypercallia)	Brazil (Pa)
32	**calidaria** (Meyrick, 1921) (Hypercallia)	Brazil (Am)
33	**chaldaica** Meyrick, 1913	Argentina
34	**colpodes** (Walsingham, 1912) (Hyphypena)	Panama
35	**complicata** Clarke, 1951	Brazil (SC)
36	**cyathopa** Meyrick, 1913	Colombia
37	**cyathopoides** Clarke, 1951	Ecuador
38	**elena** Clarke, 1951	Brazil (Am)
39	**fenestrella** Zeller, 1863	Venezuela
40	**gioia** Clarke, 1951	Bolivia
41	**margaritacea** (Meyrick, 1924) (Hypercallia)	Bolivia
42	**nigriplaga** Dognin, 1904	Colombia
	prominula Meyrick, 1913	Colombia
43	**pecten** Clarke, 1951	Guatemala
44	**perseaphaga** Clarke, 1951	Costa Rica
45	**terminalis** Clarke, 1951	Mexico

COSTOMA Busck, 1914
Phalarotarsa Meyrick, 1924

46	**basirosella** Busck, 1914	Panama
47	**cirrhophaea** (Meyrick, 1924) (Phalarotarsa)	Bolivia
48	**flavicosta** (R. Felder & Rogenhofer, 1875) (Cryptolechia)	Brazil (Am)

CRYPTOLECHIA Zeller, 1852

49	**asemanta** Dognin, 1905	Ecuador
50	**citrodeta** Meyrick, 1921	Brazil (Pa)
51	**diplosticha** Meyrick, 1926	Colombia
52	**epidesma** Walsingham, 1912	Mexico (Tab)
53	vallifera Meyrick, 1914	Guyana
53	**eucharistis** Meyrick, 1931	Argentina
54	**glischrodes** Meyrick, 1931	Argentina
55	**holopyrrha** Meyrick, 1912	Colombia
56	**hydara** Walsingham, 1912	Guatemala
57	**ichnitis** Meyrick, 1918	French Guiana
58	**microglyptis** Meyrick, 1936	Colombia
59	**pateropa** Meyrick, 1931	Brazil (SP)
60	**pentathlopa** Meyrick, 1933	Brazil (SC)
61	**percnocoma** Meyrick, 1930	Brazil (RJ)
62	**praevecta** Meyrick, 1929	Colombia
63	**remotella** (Staudinger, 1899) (Depressaria)	Argentina
64	**sciodeta** Meyrick, 1930	Brazil (RJ)
65	**semibrunnea** Dognin, 1905	Colombia
66	**taphrocopa** Meyrick, 1926	Colombia
67	**transfossa** Meyrick, 1926	Peru
68	**veniflua** Meyrick, 1914	Colombia

DOINA Clarke, 1978

69	**annulata** Clarke, 1978	Chile
70	**asperula** Clarke, 1978	Chile
71	**edmondsii** (Butler, 1883) (Depressaria)	Chile
72	**eremnogramma** Clarke, 1978	Chile
73	**flinti** Clarke, 1978	Chile
74	**glebula** Clarke, 1978	Chile
75	**inconspicua** Clarke, 1978	Chile
76	**increta** (Butler, 1883) (Orthotelia)	Chile
77	**lagneia** Clarke, 1978	Chile
78	**paralagneia** Clarke, 1978	Chile
79	**phaobregna** Clarke, 1978	Chile
80	**scariphista** (Meyrick, 1931) (Cryptolechia)	Chile
81	**subicula** Clarke, 1978	Chile
82	**trachycantha** Clarke, 1978	Chile
83	**truncata** Clarke, 1978	Chile

DOSHIA Clarke, 1978

84	**miltopeza** Clarke, 1978	Chile

ECTAGA Walsingham, 1912

85	**canescens** Walsingham, 1912	Mexico (Gue)
86	**lenta** Clarke, 1956	Argentina
87	**lictor** Walsingham, 1912	Guatemala
88	**promeces** Walsingham, 1912	Guatemala

ERITHYMA Meyrick, 1914

89	**cyanoplecta** Meyrick, 1914	Guyana

90 **trabeella** (R. Felder & Rogenhofer, 1875) (Oecophora) Brazil (Am)
polychroma Meyrick, 1914 Guyana
trabeata Meyrick, 1922, missp.

EUPRAGIA Walsingham, 1911

91 **oxinopa** Meyrick, 1929 Colombia
92 **solida** Walsingham, 1911 Mexico (Tab)

FILINOTA Busck, 1911
Lupercalia Busck, 1912
Mnesichara Walsingham, 1912 [45]

93 **brunniceps** (R. Felder & Rogenhofer, 1875) (Lecithocera) Colombia
94 **cassiteranthes** Meyrick, 1932 Bolivia
95 **dyctiota** (Walsingham, 1912) (Mnesichara) [45] Guatemala
96 **gratiosa** (R. Felder & Rogenhofer, 1875) (Lecithocera) Colombia
peruviella Busck, 1911 Peru
97 **hermosella** Busck, 1911 French Guiana
98 **ignita** (Busck, 1912) (Lupercalia) Panama
99 **ithymetra** Meyrick, 1926 [45] Colombia
100 **lamprocosma** Meyrick, 1916 French Guiana
101 **regifica** Meyrick, 1921 Brazil (Am)
102 **rhodograpta** Meyrick, 1915 Guyana
103 **sphenoplecta** Meyrick, 1921 Brazil (Am)
104 **vociferans** Meyrick, 1930 Brazil (RJ)

GNATHOTONA Meyrick, 1931

105 **thermopsamma** Meyrick, 1931 Paraguay

GONADA Busck, 1911

106 **cabima** Busck, 1912 Panama
107 **falculinella** Busck, 1911 French Guiana
108 **flavidorsis** Meyrick, 1930 Brazil (RJ)
109 **phosphorodes** Meyrick, 1922 French Guiana
110 **pyronota** Meyrick, 1924 Peru
111 **rubens** Meyrick, 1916 French Guiana

GONIONOTA Zeller, 1877

112 **acrocosma** (Meyrick, 1912) (Coptotelia) Colombia
113 **aethographa** Clarke, 1971 Costa Rica
114 **aethoptera** Clarke, 1971 Venezuela
115 **amauroptera** Clarke, 1971 Argentina
116 **amphicrena** (Meyrick, 1912) (Coptotelia) Colombia
117 **anelicta** (Meyrick, 1926) (Hypercallia) Bolivia
118 **anisodes** (Meyrick, 1916) (Hypercallia) French Guiana
119 **argopleura** Clarke, 1971 Brazil (SC)
120 **autocrena** (Meyrick, 1930) (Hypercallia) Brazil (RJ)
121 **borquiniella** (Köhler, 1940) (Hypercallia) Argentina
122 **bourquini** Clarke, 1964 Brazil (SC)
123 **captans** (Meyrick, 1931) (Hypercallia) Brazil (SP)
124 **citronota** (Meyrick, 1932) (Hypercallia) Brazil (SC)
125 **cologramma** Clarke, 1971 Venezuela
126 **comastis** Meyrick, 1909 Peru
127 **confinella** (R. Felder & Rogenhofer, 1875) (Hypercallia) Colombia
128 **constellata** (Meyrick, 1912) (Coptotelia) Colombia
129 **contrasta** Clarke, 1964 Peru
130 **cristata** Walsingham, 1912 Panama
131 **cyanaspis** (Meyrick, 1909) (Doleromima) Peru
132 **determinata** Clarke, 1964 Guyana
133 **dissita** Clarke, 1964 Trinidad
134 **dryodesma** (Meyrick, 1916) (Hypercallia) French Guiana
dryocrypta (Meyrick, 1931) (Hypercallia) Guatemala
135 **eremia** Clarke, 1971 French Guiana
136 **erotopis** (Meyrick, 1926) (Hypercallia) Bolivia
137 **erythroleuca** (Meyrick, 1928) (Hypercallia) Peru
138 **eurydryas** (Meyrick, 1926) (Hypercallia) Colombia
139 **euthyrsa** (Meyrick, 1930) (Hypercallia) Ecuador
140 **excavata** Clarke, 1964 Mexico (Ver)
141 **extima** Clarke, 1964 Costa Rica
142 **festicola** (Meyrick, 1924) (Hypercallia) Peru
143 **fimbriata** Clarke, 1964 Panama
144 **gaiophanes** Clarke, 1971 Brazil (SC)
145 **habristis** (Meyrick, 1914) (Hypercallia) Guyana
146 **hydrogramma** (Meyrick, 1912) (Coptotelia) Colombia
147 **hypoleuca** Clarke, 1971 Venezuela
148 **hyptiotes** Clarke, 1964 Mexico (Ver)
149 **incalescens** (Meyrick, 1914) (Hypercallia) Colombia
150 **incisa** Meyrick, 1909 Bolivia
151 **incontigua** Clarke, 1964 Venezuela
152 **insignata** Clarke, 1971 Ecuador
153 **insulana** Clarke, 1968 Dominica
154 **intonans** (Meyrick, 1933) (Hypercallia) Argentina
155 **ioleuca** (Meyrick, 1912) (Coptotelia) Argentina
156 **isastra** (Meyrick, 1926) (Hypercallia) Colombia
157 **isodryas** (Meyrick, 1921) (Hypercallia) Brazil (Pa)
158 **isophylla** Meyrick, 1909 Peru
159 **lecithitis** (Meyrick, 1912) (Coptotelia) Argentina
160 **leucoporpa** (Meyrick, 1926) (Hypercallia) Colombia
161 **lichenista** (Meyrick, 1926) (Hypercallia) Colombia
162 **luteola** (R. Felder & Rogenhofer, 1875) (Carcina) Brazil (Am)
163 **melobaphes** Walsingham, 1912 Panama
164 **menura** Clarke, 1971 Panama
165 **militaris** (Meyrick, 1914) (Hypercallia) Colombia
166 **mimulina** (Butler, 1883) (Agriocoma) Chile
araucana (Bartlett-Calvert, 1893) (Agriocoma) Chile
167 **mitis** (Meyrick, 1914) (Hypercallia) Peru
168 **notodontella** Zeller, 1877 Colombia
169 **oligarcha** (Meyrick, 1913) (Coptotelia) Peru
170 **oriphanta** (Meyrick, 1928) (Hypercallia) Colombia
171 **oxybela** Clarke, 1971 Peru
172 **paravexillata** Clarke, 1971 Venezuela
173 **periphereia** Clarke, 1964 Ecuador
174 **persistis** (Meyrick, 1914) (Hypercallia) Peru
175 **phocodes** Meyrick, 1909 Peru
176 **phthiochroma** Clarke, 1971 Bolivia
177 **pialea** (Meyrick, 1921) (Hypercallia) Brazil (Pa)
178 **poecilia** Clarke, 1971 Venezuela
179 **praeclivis** (Meyrick, 1921) (Hypercallia) Peru
180 **prolectans** (Meyrick, 1926) (Hypercallia) Colombia
181 **pyrocausta** (Meyrick, 1931) (Hypercallia) Colombia
182 **pyrrhotrota** (Meyrick, 1932) (Hypercallia) Bolivia
183 **rhacina** Walsingham, 1912 Guatemala
184 **rosacea** (Forbes, 1931) (Hypercallia) Haiti
185 **satrapis** (Meyrick, 1914) (Hypercallia) Colombia
186 **saulopis** Meyrick, 1909 Peru
187 **selene** Clarke, 1971 Brazil (SC)
188 **sphenogramma** Clarke, 1971 Venezuela
189 **teganitis** Meyrick, 1909 Peru
190 **tenebralis** (Hampson, 1906) (Salobrena) Brazil (RJ)
charagma Clarke, 1971 Brazil (SC)
191 **transversa** Clarke, 1971 Brazil (Pr)
192 **uberrima** (Meyrick, 1914) (Hypercallia) Peru
193 **vexillata** (Meyrick, 1913) (Coptotelia) Peru
194 **vivida** (Meyrick, 1924) (Hypercallia) Bolivia

HABROPHYLAX Meyrick, 1931

195 **chalcochtha** Meyrick, 1931 Brazil (SP)

HAMADERA Busck, 1914

196 **aurea** Busck, 1914 Panama

HASTAMEA Fletcher, 1929, repl. name
Hasta Busck, 1911, preocc. (Kirkaldy, 1906 [Hemiptera])

197 **argentidorsella** (Busck, 1911) (Hasta) Brazil (Pr)

HIMOTICA Meyrick, 1912

198 **thyrsitis** Meyrick, 1912 Brazil (SP)

HYPERCALLIA Stephens, 1834
Agriocoma Zeller, 1877
Brachyplatea Zeller, 1877
Eumimographe Dognin, 1905 [19]

199 **alexandra** (Meyrick, 1909) (Gonionota) Peru
200 **argyropa** Meyrick, 1914 Peru
201 **arista** Walsingham, 1912 Mexico (Ver)
202 **bruneri** Busck, [1934] Cuba
203 **catenella** (Zeller, 1877) (Agriocoma) Peru
204 **chionastra** Meyrick, 1926 Colombia
205 **chionopis** Meyrick, 1916 French Guiana
206 **citroclista** Meyrick, 1930 Brazil (RJ)
207 **cnephaea** (Walsingham, 1912) (Gonionota) Panama
208 **crocatella** Zeller, 1877 Colombia
209 **cupreata** (Dognin, 1905) (Eumimographe) Ecuador
210 **diplotrocha** Meyrick, 1937 Argentina
211 **gnorisma** (Walsingham, 1912) (Coptotelia) Mexico (Ver)
212 **halobapta** Meyrick, 1930 Brazil (RJ)
213 **heliodepta** Meyrick, 1932 Mexico
214 **heliomima** Meyrick, 1930 Colombia
215 **heterochroma** Clarke, 1971 Venezuela
216 **incensella** (Zeller, 1877) (Brachyplatea) Peru
217 **longimaculata** (Dognin, 1905) (Eumimographe) Ecuador
218 **loxochorda** Meyrick, 1926 Colombia
219 **lydia** (Druce, 1901) (Atteria), **n. comb.** [20] Ecuador
220 **miltopa** (Meyrick, 1912) (Coptotelia) Colombia
tunicata (Busck, 1914) (Cryptolechia) Panama
221 **miniata** (Dognin, 1905) (Cryptolechia) Ecuador
222 **niphocycla** Meyrick, 1926 Colombia
223 **obliquistriga** Dognin, 1905 Ecuador
224 **orthochaeta** (Meyrick, 1913) (Coptotelia) Peru
225 **phlebotes** (Walsingham, 1912) (Gonionota) Guatemala
226 **psittacopa** (Meyrick, 1912) (Coptotelia) Colombia
227 **rhodosarca** (Walsingham, 1912) (Cryptolechia) Guatemala
rhodosarea Meyrick, 1922, missp.
228 **syntoma** (Walsingham, 1912) (Hyphypena) Panama

IDIOCRATES Meyrick, 1909

229 **balanitis** Meyrick, 1909 Bolivia

LUCYNA Clarke, 1978

230 **fenestella** (Zeller, 1874) (Cryptolechia) Chile
thyridopa (Meyrick, 1912) (Coptotelia), repl. name Chile

MACHIMIA Clemens, 1860

231 **aethostola** Meyrick, 1931 Brazil (SP)
232 **anthracospora** Meyrick, 1934 Brazil (SC)
233 **caduca** (Walsingham, 1912) (Cryptolechia) Guatemala
234 **chorrera** (Busck, 1914) (Cryptolechia) Panama
235 **cruda** Meyrick, 1926 Colombia
236 **desertorum** (Berg, 1875) (Cryptolechia) Argentina
237 **diagrapha** Meyrick, 1931 Paraguay
238 **dolopis** (Walsingham, 1912) (Cryptolechia) Mexico (Gue)
239 **eothina** Meyrick, 1920 French Guiana
240 **ignicolor** (Busck, 1914) (Cryptolechia) Panama
241 **illuminella** (Busck, 1914) (Cryptolechia) Panama
241.1 **intaminata** Meyrick, 1922 Brazil
242 **morata** Meyrick, 1912 Argentina
243 **neuroscia** Meyrick, 1930 Brazil (Pa)
244 **notella** (Busck, 1914) (Cryptolechia) Panama
245 **oxybela** Meyrick, 1931 Brazil
246 **peperita** (Walsingham, 1912) (Cryptolechia) Guatemala
247 **perianthes** Meyrick, 1922 French Guiana
248 **pyrocalyx** Meyrick, 1922 Brazil (SC)
249 **pyrograpta** Meyrick, 1932 Bolivia
250 **rogifera** Meyrick, 1914 Guyana
251 **sejunctella** (Walker, 1864) (Gelechia) Brazil (Am)
252 **trigama** (Meyrick, 1928) (Cryptolechia) USA (Tx)
253 **trunca** Meyrick, 1930 Brazil (Pa)

MAESARA Clarke, 1968

254 **gallegoi** Clarke, 1968 Colombia

MELANEULIA Butler, 1883

255 **hecate** Butler, 1883 Chile

MUNA Clarke, 1978

256 **zostera** Clarke, 1978 Chile

NEDENIA Clarke, 1978

257 **rhodochra** Clarke, 1978 Chile

NEMATOCHARES Meyrick, 1931

258 **citraulax** Meyrick, 1931 Brazil

OSMARINA Clarke, 1978

259 **argilla** Clarke, 1978 Chile

PALINORSA Meyrick, 1924

260 **acritomorpha** Clarke, 1964 Peru
261 **literatella** (Busck, 1911) (Pleurota) French Guiana
262 **raptans** (Meyrick, 1920) (Orsotricha) Peru
263 **zonaria** Clarke, 1964 Bolivia

PERZELIA Clarke, 1978

264 **arda** Clarke, 1978 Chile

PHILTRONOMA Meyrick, 1914

265 **roseicorpus** (Dognin, 1910) (Oecophora) French Guiana

PHOLCOBATES Meyrick, 1931

266 **flagelliformis** Meyrick, 1931 Brazil (ES)

PHYTOMIMIA Walsingham, 1912

267 **chlorophylla** Walsingham, 1912 Guatemala
redundans Walsingham, 1912 Guatemala
silvicolor Meyrick, 1932 Panama
268 **cynegetis** Meyrick, 1932 Peru
269 **pyrrhophthalma** Meyrick, 1932 Costa Rica

PISINIDEA Butler, 1883

270 **viridis** Butler, 1883 Chile

PROFILINOTA Clarke, 1973

271 **phillita** Clarke, 1973 Venezuela

PSEUDOCENTRIS Meyrick, 1921

272 **testudinea** Meyrick, 1921 Peru

PSILOCORSIS Clemens, 1860
Paepia Walker, 1864
Hagno Chambers, 1872

273 **argyropasta** Walsingham, 1912 Mexico (Gue)
274 **carpocapsella** (Walker, 1864) (Paepia) Brazil (Am)
275 **exagitata** (Meyrick, 1926) (Cryptolechia) Colombia
276 **indalma** Walsingham, 1912 Guatemala
277 **melanophthalma** (Meyrick, 1929) (Cryptolechia) Colombia
278 **minerva** (Meyrick, 1928) (Cryptolechia) Colombia
279 **propriella** Zeller, 1877 Colombia
280 **purpurascens** Walsingham, 1912 Guatemala

PSITTACASTIS Meyrick, 1909
Necedes Walsingham, 1912

281 **argentata** Meyrick, 1921 Brazil (Am)
282 **championella** (Walsingham, 1912) (Necedes) Guatemala
283 **cocae** (Busck, 1931) (Eucleodora) Peru
284 **cosmodoxa** Meyrick, 1921 Peru
285 **eumolybda** Meyrick, 1926 Peru
286 **eurychrysa** Meyrick, 1909 Bolivia
287 **gaulica** Meyrick, 1909 Bolivia
288 **incisa** (Walsingham, 1912) (Necedes) Mexico (Tab)
289 **molybdaspis** Meyrick, 1926 Peru
290 **pictrix** Meyrick, 1921 Colombia
291 **propriella** (Walker, 1864) (Gelechia) Brazil (Am)
292 **pyrsophanes** Meyrick, 1936 Peru
293 **stigmaphylli** (Walsingham, 1912) (Necedes) Jamaica
brevipalpis (Walsingham, 1912) (Necedes) Jamaica
294 **superatella** (Walker, 1864) (Gelechia) Brazil (Am)
ambigua (R. Felder & Rogenhofer, 1875) (Oecophora) Brazil (Am)
295 **trierica** Meyrick, 1909 Bolivia

RHINDOMA Busck, 1914

296 **rosapicella** Busck, 1914 Panama

SCOLIOGRAPHA Meyrick, 1916

297 **argospila** Meyrick, 1916 French Guiana

TARUDA Walker, 1864
Ecliptoloma Zeller, 1877

298 **apicella** Walker, 1864 Brazil (Am)
299 **cuneatella** Walker, 1864 Brazil (Am)
obydella (R. Felder & Rogenhofer, 1875) (Rhinosia) Brazil (Am)
300 **haemoplecta** Meyrick, 1931 Brazil (ES)
301 **hemiommata** (Zeller, 1877) (Ecliptoloma) Brazil
302 **leucochna** Meyrick, 1921 Peru
303 **oblitella** (R. Felder & Rogenhofer, 1875) (Lecithocera) Brazil (Am)

TRYCHERODES Meyrick, 1914
Teratomorpha Walsingham, 1912, preocc. (Turner, 1896 [Oecophoridae])

304 **albifrons** (Walsingham, 1912) (Teratomorpha) Mexico (Tab)
305 **chilibrella** (Busck, 1914) (Teratomorpha) Panama
306 **producta** (Walsingham, 1912) (Teratomorpha) Guatemala

ZEMIOCRITA Meyrick, 1933

307 **spermatopis** Meyrick, 1933 Argentina

Ethmiinae

ERYSIPTILA Meyrick, 1914 [21]
Cyrictodes Meyrick, 1926

308 **clevelandi** (Busck, 1914) (Borkhausenia) Panama
phormophora (Meyrick, 1926) (Cyrictodes) Costa Rica

ETHMIA Hübner, [1819]
Psecadia Hübner, [1825]
Anesychia Hübner, [1825]
Disthymnia Hübner, [1825]
Melanoleuca Stephens, 1829
Aedia Duponchel, 1836
Chalybe Duponchel, 1836
Azinis Walker, 1863
Tamarrha Walker, 1864
Ceratophysetis Meyrick, 1887
Theoxenia Walsingham, 1887
Babaiaxa Busck, 1902
Wiltshireia Amsel, 1949

309 **abraxasella** (Walker, 1864) (Psecadia)
a) **abraxasella** (Walker, 1864) (Psecadia) Dom. Rep.
aureoapicella (Möschler, 1890) (Psecadia) Puerto Rico
abraxella Meyrick, 1914, missp.

	b) **clarissa** Busck, 1914	Cuba
310	**albicostella** (Beutenmüller, 1889) (Psecadia)	USA (Co)
311	**angustalatella** Powell, 1973	Mexico (NL)
312	**arctostaphylella** (Walsingham, 1880) (Psecadia)	USA (Ca)
	obscurella (Beutenmüller, 1888) (Psecadia)	USA (Ca)
	mediella Busck, 1913	USA (Ca)
313	**baja** Powell, 1973	Mexico (BCS)
314	**baliostola** Walsingham, 1912	Costa Rica
	baliostoma Busck, 1914, missp.	
315	**bittenella** (Busck, 1906) (Tamarrha)	USA (Tx)
316	**calumniella** Powell, 1973	Brazil (Pa)
317	**catapeltica** Meyrick, 1924	Costa Rica
318	**cellicoma** Meyrick, 1931	Paraguay
319	**chalcodora** Meyrick, 1912	Argentina
320	**chalcogramma** Powell, 1973	Bolivia
321	**chemsaki** Powell, 1959	Costa Rica
322	**clarkei** Powell, 1973	Mexico (QR)
323	**clava** Powell, 1973	Mexico (Ver)
324	**confusella** (Walker, 1863) (Hyponomeuta)	Dom. Rep.
	strigosella (Walker, 1864) (Cryptolechia)	Dom. Rep.
	ingricella (Möschler, 1890) (Hyponomeuta)	Puerto Rico
	strigosa (Cockerell, 1891) (Psecadia), missp.	
325	**confusellastra** Powell, 1973	Mexico (Yuc)
326	**conglobata** Meyrick, 1912	Colombia
*327	**coquillettella** Busck, 1907	USA (Ca)
328	**cordia** Powell, 1973	Mexico (Yuc)
329	**coronata** Walsingham, 1912	Mexico (Gue)
	abdominella Busck, 1912	Mexico (Gue)
330	**cubensis** Busck, [1934]	Cuba
331	**cupreonivella** (Walsingham, 1880) (Psecadia)	Brazil (ES)
332	**cyanea** Walsingham, 1912	Mexico (Ver)
333	**cypraeella** (Zeller, 1863) (Psecadia)	Venezuela
334	**cypraspis** Meyrick, 1930	Brazil (Pa)
335	**davisella** Powell, 1973	Mexico (Tam)
336	**delliella** (Fernald, 1891) (Psecadia)	USA (Tx)
337	**discostrigella** (Chambers, 1877) (Anesychia)	
	a) **discostrigella** (Chambers, 1877) (Anesychia)	USA (Co)
	b) **subcaerulea** (Walsingham, 1880) (Psecadia)	USA (Ca)
338	**duckworthi** Powell, 1973	Panama
339	**elutella** Busck, 1914	Panama
340	**epilygella** Powell, 1973	Brazil (SC)
341	**exornata** (Zeller, 1877) (Psecadia)	Peru
	exornatella Busck, 1906, missp.	
342	**farrella** Powell, 1973	Jamaica
343	**festiva** Busck, 1914	Panama
	xantholitha Meyrick, 1929	Colombia
344	**flavicaudata** Walsingham, 1912	Mexico (Ver)
345	**fritillella** Powell, 1973	Brazil (SC)
346	**gelidella** (Walker, 1864) (Tamarrha)	Jamaica
347	**gigantea** Busck, 1914	Mexico (Ver)
348	**hagenella** (Chambers, 1878) (Anesychia)	
	a) **hagenella** (Chambers, 1878) (Anesychia)	USA (Tx)
	b) **josephinella** Dyar, 1902	USA (NM)
349	**hammella** Busck, 1910	Costa Rica
350	**heptastica** Walsingham, 1912	Mexico (Gue)
351	**hieroglyphica** Powell, 1973	Bolivia
352	**hiramella** Busck, 1914	Cuba
353	**hodgesella** Powell, 1973	USA (Az)
354	**howdeni** Powell, 1973	Mexico (Sin)
355	**humilis** Powell, 1973	Jamaica
356	**iridella** Powell, 1973	Mexico (Pue)
357	**janzeni** Powell, 1973	Mexico (Oax)
358	**joviella** Walsingham, 1897	Grenada
359	**julia** Powell, 1973	Puerto Rico
360	**kirbyi** (Möschler, 1890) (Psecadia)	Puerto Rico
361	**lichyi** Powell, 1973	Venezuela
362	**linda** Busck, 1914	Venezuela
363	**linsdalei** Powell, 1973	Mexico (Oax)
364	**mansita** Busck, 1914	Mexico (Pue)
365	**marmorea** (Walsingham, 1888) (Psecadia)	USA (Az)
366	**mnesicosma** Meyrick, 1924	Costa Rica
367	**mulleri** Busck, 1910	Mexico (Pue)
	mülleri Busck, 1912, missp. (incorr. spell.)	
368	**nigritaenia** Powell, 1973	Mexico (Yuc)
369	**nivosella** (Walker, 1864) (Tamarrha)	Jamaica
	adustella (Zeller, 1877) (Psecadia)	Puerto Rico
	niveosella (Busck, 1906) (Tamarrha), missp.	
370	**notatella** (Walker, 1863) (Psecadia)	Dom. Rep.
	xanthorrhoa (Zeller, 1877) (Psecadia)	Puerto Rico
371	**notomurinella** Powell, 1973	Argentina
372	**omega** Powell, 1973	Brazil (RS)
373	**oterosella** Busck, [1934]	Cuba
374	**pala** Powell, 1973	Mexico (Sin)
375	**papiella** Powell, 1973	Mexico (Sin)
376	**paucella** (Walker, 1863) (Hyponomeuta)	Dom. Rep.
377	**penthica** Walsingham, 1912	Mexico (Oax)
378	**perpulchra** Walsingham, 1912	Mexico (Ver)
379	**phoenicura** Meyrick, 1932	Mexico (BCS)
380	**phylacis** Walsingham, 1912	
	a) **phylacis** Walsingham, 1912	Mexico [Sin]
	b) **ornata** Busck, [1934]	Cuba
381	**phylacops** Powell, 1973	Mexico (Yuc)
382	**piperella** Powell, 1973	Jamaica
383	**plaumanni** Powell, 1973	Brazil (SC)
384	**playa** Powell, 1973	Mexico (Sin)
385	**prattiella** Busck, 1915	USA (Tx)
386	**proximella** Busck, 1912	Mexico (Pue)
387	**punctessa** Powell, 1973	Mexico (NL)
388	**sandra** Powell, 1973	El Salvador
389	**scutula** Powell, 1973	Mexico (Sin)
390	**scythropa** Walsingham, 1912	Costa Rica
391	**semilugens** (Zeller, 1872) (Psecadia)	USA (Tx)
	multipunctella (Chambers, 1874) (Anesychia)	USA (Tx)
	semiopaca (Grote, 1881) (Psecadia)	USA (Co)
	plumbeella (Beutenmüller, 1889) (Psecadia)	USA (Tx)
392	**semiombra** Dyar, 1902	
	a) **semiombra** Dyar, 1902	USA (Tx)
	b) **nebulombra** Powell, 1973	Mexico (Yuc)
393	**semitenebrella** Dyar, 1902	USA (Az)
394	**similatella** Busck, 1920	Guatemala
395	**sphenisca** Powell, 1973	Mexico (Dur)
396	**striatella** Busck, 1913	Mexico (Pue)
397	**submissa** Busck, 1914	Cuba
398	**subnigritaenia** Powell, 1973	Mexico (DF)
399	**subsimilis** Walsingham, 1897	Jamaica
400	**terpnota** Walsingham, 1912	Costa Rica
401	**transversella** Busck, 1914	Costa Rica
402	**ungulatella** Busck, 1914	Panama
403	**volcanella** Powell, 1973	Mexico (Oax)
404	**wellingi** Powell, 1973	Mexico (Yuc)
405	**zebrata** Powell, 1959	Mexico (Pue)

MACROCIRCA Meyrick, 1931 [22]

406	**strabo** Meyrick, 1931	Argentina

Peleopodinae

DURRANTIA Busck, 1908
Dolidiria Busck, 1912

407	**acompsa** Walsingham, 1912	Panama
	monotona (Amsel, 1956) (Stenoma)	Venezuela
408	**amabilis** Walsingham, 1912	Guatemala
409	**arcanella** (Busck, 1912) (Dolidiria)	Panama
410	**flaccescens** Meyrick, 1925	Peru
411	**pugnax** Walsingham, 1912	Guatemala
412	**resurgens** Walsingham, 1912	Guatemala

PELEOPODA Zeller, 1877
Theatria Walsingham, 1912

413	**convoluta** Duckworth, 1970	Venezuela
414	**lobitarsis** Zeller, 1877	Panama
	lobitarsus Hodges, 1974, missp.	
415	**navigatrix** (Meyrick, 1912) (Xylorycta)	Colombia
416	**semocrossa** Meyrick, 1930	Bolivia
417	**spudasma** (Walsingham, 1912) (Theatria)	Panama

PSEUDEROTIS Clarke, 1956

418	**cannescens** Clarke, 1956	Argentina
419	**thamnolopha** (Meyrick, 1932) (Asapharca)	Costa Rica

SCHISTONOEA Forbes, 1931

420	**fulvidella** (Walsingham, 1897) (Brachmia)	Virgin Is. (St. Thomas)

Stenomatinae

ANADASMUS Walsingham, 1897

421	**accurata** (Meyrick, 1916) (Stenoma)	French Guiana
	perjura (Meyrick, 1925) (Stenoma)	Brazil (Am)
422	**anceps** (Butler, 1877) (Cryptolechia)	Brazil (Am)
	praeceps (Meyrick, 1915) (Stenoma)	French Guiana
423	**arenosa** (Meyrick, 1916) (Stenoma)	French Guiana
424	**byrsinites** (Meyrick, 1912) (Stenoma)	Colombia
425	**caliginea** (Meyrick, 1930) (Stenoma)	Brazil (SP)
	lianthes (Meyrick, 1932) (Stenoma)	Brazil (SC)
426	**capnocrossa** (Meyrick, 1925) (Stenoma)	Brazil (Am)
427	**chlorothrota** (Meyrick, 1932) (Stenoma)	Bolivia
428	**endochra** (Meyrick, 1925) (Stenoma)	Brazil (Am)
429	**germinans** (Meyrick, 1925) (Stenoma)	Colombia
430	**incitatrix** (Meyrick, 1925) (Stenoma)	Argentina
431	**ischioptila** (Meyrick, 1925) (Stenoma)	Colombia
432	**leontodes** (Meyrick, 1915) (Gonioterma)	Surinam
433	**lithogypsa** (Meyrick, 1932) (Stenoma)	Brazil (Pa)
434	**nonagriella** (Walker, 1864) (Cryptolechia)	Brazil (Am)
435	**obmutescens** (Meyrick, 1916) (Stenoma)	French Guiana
	pleutotricha (Meyrick, 1925) (Stenoma)	Brazil (Am)
436	**paurocentra** (Meyrick, 1912) (Stenoma)	Colombia
437	**pelinitis** (Meyrick, 1912) (Stenoma)	Colombia
438	**pelodes** (Walsingham, 1913) (Stenoma)	Panama
	scortea (Meyrick, 1915) (Stenoma)	Brazil (Am)
439	**plebicola** (Meyrick, 1918) (Stenoma)	French Guiana
440	**quadratella** (Walker, 1864) (Cryptolechia)	Brazil (Pa)
	hebes (Dognin, 1905) (Mesoptycha)	Colombia

gerda (Busck, 1914) (Gonioterma) Panama
441 **sororia** (Zeller, 1877) (Cryptolechia) Colombia
catapsecta (Meyrick, 1915) (Stenoma) Guyana
442 **vacans** (Meyrick, 1916) (Stenoma) French Guiana
443 **venosella** (Walker, 1864) (Cryptolechia) Brazil (Am)
pleximorpha (Meyrick, 1930) (Stenoma) Bolivia
dictyogramma (Meyrick, 1932) (Stenoma) Brazil (Am)

ANAPATRIS Meyrick, 1932

444 **chersopsamma** Meyrick, 1932 Panama

ANTAEOTRICHA Zeller, 1854
Mesoptycha Zeller, 1854
Brachiloma Clemens, 1863
Harpalyce Chambers, 1874
Ide Chambers, 1880
Antoeotricha Walsingham, 1881, missp.
Aedemoses Walsingham, 1912
Athleta Walsingham, 1912
Prasolithites Meyrick, 1912
Aphanoxena Meyrick, 1915
Psephomeres Meyrick, 1916
Eumiturga Meyrick, 1925

445 **acrograpta** (Meyrick, 1915) (Aphanoxena) Guyana
446 **acronephela** Meyrick, 1915 Guyana
447 **actista** (Meyrick, 1913) (Stenoma) Venezuela
448 **addon** (Busck, 1911) (Stenoma) Guyana
cicadella (Sepp, [1852]-55) (Phalaena), preocc. (not Sepp, [1832-40]) Surinam
449 **adductella** (Walker, 1864) (Cryptolechia) Brazil (Am)
450 **adjunctella** (Walker, 1864) (Cryptolechia) Brazil (Am)
additella (Walker, 1864) (Cryptolechia) Brazil (Am)
absconditella (Walker, 1864) (Cryptolechia) Brazil (Am)
451 **admixta** (Walsingham, 1913) (Stenoma) Mexico (Gue)
452 **adornata** (Meyrick, 1915) (Stenoma) Peru
453 **aequabilis** (Meyrick, 1916) (Stenoma) French Guiana
454 **aerinotata** (Butler, 1877) (Cryptolechia) Brazil (Am)
sperata (Busck, 1911) (Stenoma) French Guiana
455 **affinis** R. Felder & Rogenhofer, 1875 Brazil (Am)
tanysta Meyrick, 1915 Guyana
anticharis (Meyrick, 1916) (Stenoma) French Guiana
456 **aggravata** (Meyrick, 1916) (Stenoma) French Guiana
457 **aglypta** Meyrick, 1925 Brazil (Am)
458 **albicilla** (Zeller, 1854) (Cryptolechia) Venezuela
albicella Busck, [1934], missp.
459 **albifrons** Zeller, 1877 Brazil
460 **albilimbella** (R. Felder & Rogenhofer, 1875) (Gelechia) Brazil (Am)
nuntia (Meyrick, 1925) (Stenoma) Peru
461 **albitincta** (Meyrick, 1930) (Stenoma) Brazil (ES)
pauroconis (Meyrick, 1932) (Stenoma) Brazil (SC)
462 **albovenosa** Zeller, 1877 Peru
463 **amicula** Zeller, 1877 Panama
464 **ammodes** (Walsingham, 1913) (Stenoma) Mexico (Tab)
465 **amphilyta** Meyrick, 1916 French Guiana
466 **amphizyga** Meyrick, 1930 Brazil (Pa)
467 **anaclintris** Meyrick, 1916 French Guiana
468 **arachnia** (Meyrick, 1915) (Stenoma) Guyana
469 **aratella** (Walker, 1864) (Cryptolechia) Brazil (Am)
470 **argocorys** (Meyrick, 1931) (Stenoma) Brazil (ES)
471 **arystis** Meyrick, 1915 Guyana
472 **assecta** Zeller, 1877 Peru
473 **astynoma** (Meyrick, 1915) (Aphanoxena) Guyana
474 **atmospora** (Meyrick, 1925) (Stenoma) Colombia
475 **axena** Meyrick, 1916 French Guiana
476 **balanocentra** (Meyrick, 1915) (Aphanoxena) Guyana
477 **ballista** (Meyrick, 1916) (Stenoma) French Guiana
478 **basalis** Zeller, 1854 Brazil
harpobathra Meyrick, 1916 Argentina
479 **basiferella** (Walker, 1864) (Cryptolechia) Brazil (Am)
480 **basilaris** (Busck, 1914) (Stenoma) Panama
481 **basirubrella** (Walker, 1864) (Cryptolechia) Brazil (Am)
482 **bathrotoma** (Meyrick, 1925) (Stenoma) Brazil (Am)
483 **biarcuata** Meyrick, 1926 Colombia
stenobathra Meyrick, 1932 Panama
vogli Amsel, 1956 Venezuela
484 **bicolor** (Zeller, 1839) (Stenoma) Brazil
dissimilis Kearfott, 1911 Brazil (SP)
annixa Meyrick, 1918 Brazil (SP)
485 **bilinguis** (Meyrick, 1918) (Stenoma) French Guiana
486 **binubila** Zeller, 1854 Brazil (Pa)
aporodes Meyrick, 1915 Surinam
487 **bipupillata** Meyrick, 1930 Brazil (Am)/Paraguay
488 **bracatingae** (Köhler, 1943) (Stenoma) Argentina
489 **brachysaris** Meyrick, 1916 French Guiana
490 **brochota** Meyrick, 1915 Peru
491 **caenochytis** (Meyrick, 1915) (Stenoma) Guyana
492 **camarina** Meyrick, 1915 Guyana
493 **campylodes** Meyrick, 1916 French Guiana
494 **cantharitis** (Meyrick, 1916) (Aphanoxena) French Guiana
495 **caprimulga** (Walsingham, 1912) (Stenoma) Mexico (Ver)
496 **capsiformis** (Meyrick, 1930) (Stenoma) Brazil (Ba)
497 **capsulata** Meyrick, 1918 French Guiana
498 **carabodes** (Meyrick, 1915) (Stenoma) Guyana
499 **carabophanes** Meyrick, 1932 Colombia
500 **carbasea** (Meyrick, 1915) (Stenoma) Brazil (RJ)
501 **caryograpta** (Meyrick, 1930) (Stenoma) Brazil (SP)
502 **cathagnista** Meyrick, 1925 Brazil (Am)
503 **catharactis** Meyrick, 1930 Brazil
504 **cedroxyla** Meyrick, 1930 Brazil (SP)
505 **celidotis** Meyrick, 1925 Peru
506 **ceratistes** (Walsingham, 1912) (Stenoma) Mexico (Gue)
507 **chalastis** (Meyrick, 1915) (Stenoma) Guyana
508 **chalinophanes** (Meyrick, 1931) (Stenoma) Bolivia
509 **chilosema** (Meyrick, 1918) (Stenoma) French Guiana
510 **christocoma** Meyrick, 1915 Peru
511 **cicadella** (Sepp, [1832-40]) (Phalaena) Surinam
512 **cirrhoxantha** (Meyrick, 1915) (Stenoma) French Guiana
513 **cleopatra** Meyrick, 1925 Brazil
514 **cnemosaris** (Meyrick, 1925) (Stenoma) Brazil (Pa)
515 **colposaris** (Meyrick, 1925) (Stenoma) Brazil (Am)
516 **comosa** (Walsingham, 1912) (Stenoma) Mexico (Ver)
517 **compsographa** Meyrick, 1916 French Guiana
518 **compsoneura** (Meyrick, 1925) (Stenoma) French Guiana
519 **confixella** (Walker, 1864) (Cryptolechia) Brazil (Am)
520 **congelata** Meyrick, 1926 Peru
521 **coniopa** (Meyrick, 1925) (Stenoma) Brazil (Pa)
522 **constituta** (Meyrick, 1925) (Stenoma) French Guiana
523 **constricta** (Meyrick, 1926) (Stenoma) Colombia
524 **conturbatella** (Walker, 1864) (Cryptolechia) Brazil (Am)
illucidella (Walker, 1864) (Cryptolechia) Brazil (Am)
525 **copromima** Meyrick, 1930 French Guiana
citrophaea (Meyrick, 1931) (Stenoma) Brazil (RJ)
526 **coriodes** Meyrick, 1915 Guyana
527 **corvigera** Meyrick, 1915 Guyana
528 **cosmoterma** Meyrick, 1930 Brazil (RS)
529 **costatella** (Walker, 1864) (Cryptolechia) Brazil (Am)
530 **cremastis** (Meyrick, 1925) (Stenoma) Peru
531 **cryeropis** Meyrick, 1925 Mexico (Gue)
532 **crysiphaea** (Meyrick, 1925) (Stenoma) Brazil (Pa)
533 **cyclobasis** Meyrick, 1930 Brazil (SP)
534 **cycnolopha** (Meyrick, 1925) (Stenoma) Peru
535 **cycnomorpha** Meyrick, 1925 Brazil (Pa)
536 **cymogramma** (Meyrick, 1925) (Stenoma) Peru
537 **cyprodeta** Meyrick, 1930 Brazil (RS)
538 **deltopis** Meyrick, 1915 Guyana
539 **demas** (Busck, 1911) (Stenoma) French Guiana
540 **demotica** (Walsingham, 1912) (Stenoma) Mexico (Gue)
541 **deridens** Meyrick, 1925 Bolivia
542 **desecta** (Meyrick, 1918) (Stenoma) French Guiana
543 **destillata** (Zeller, 1877) (Cryptolechia) Panama
544 **diacta** (Meyrick, 1916) (Stenoma) French Guiana
545 **diffracta** Meyrick, 1916 French Guiana
546 **diplarcha** Meyrick, 1915 Guyana
arachniotis Meyrick, 1930 Brazil (Pa)
547 **diplophaea** Meyrick, 1916 French Guiana
548 **diplosaris** (Meyrick, 1915) (Stenoma) Guyana
549 **dirempta** (Zeller, 1855) (Cryptolechia) Brazil
550 **discalis** (Busck, 1914) (Stenoma) Panama
551 **discolor** (Walsingham, 1912) (Stenoma) Guatemala
552 **disjecta** (Zeller, 1854) (Cryptolechia) Brazil
553 **dissona** (Meyrick, 1925) (Stenoma) Brazil (Am)
554 **doleopis** (Meyrick, 1915) (Stenoma) Guyana
555 **dromica** (Meyrick, 1925) (Stenoma) Brazil (Am)
556 **elaeodes** (Walsingham, 1913) (Stenoma) Mexico (Ver)
557 **elatior** (R. Felder & Rogenhofer, 1875) (Cryptolechia) Brazil (Am)
558 **encyclia** Meyrick, 1915 Colombia
559 **enodata** Meyrick, 1916 French Guiana
560 **epicrossa** (Meyrick, 1932) (Stenoma) Peru
561 **epignampta** Meyrick, 1915 Peru
562 **episimbla** (Meyrick, 1915) (Aphanoxena) Guyana
563 **ergates** (Walsingham, 1913) (Stenoma) Mexico (Tab)
564 **erotica** (Meyrick, 1916) (Stenoma) French Guiana
565 **eucoma** Meyrick, 1925 Brazil (Am)
566 **euthrinca** Meyrick, 1915 Colombia
567 **exasperata** (Meyrick, 1916) (Stenoma) French Guiana
568 **excisa** Meyrick, 1916 French Guiana
569 **extenta** (Busck, 1920) (Stenoma) Guatemala
ptilocrates Meyrick, 1932 Panama
570 **exusta** Meyrick, 1916 French Guiana
571 **falsidica** (Meyrick, 1915) (Stenoma) Surinam
572 **fascicularis** Zeller, 1854 Brazil
leprosa (R. Felder & Rogenhofer, 1875) (Cryptolechia) Brazil (Am)
gunni (Busck, 1911) (Stenoma) French Guiana
leucogramma Meyrick, 1915 Guyana
573 **fasciata** (Busck, 1911) (Stenoma) French Guiana
574 **filiferella** (Walker, 1864) (Cryptolechia) Brazil (Am)
niviliturella (Walker, 1864) (Gelechia) Brazil (Am)
menestella (Walsingham, 1913) (Stenoma) Panama
575 **flocculosa** (Meyrick, 1925) (Eumiturga) Brazil (Am)
576 **forreri** (Walsingham, 1913) (Stenoma) Mexico (Dur)
577 **fractilinea** (Walsingham, 1912) (Stenoma) Mexico (Tab)
578 **fractinubes** (Walsingham, 1912) (Stenoma) Panama
579 **fraterna** (R. Felder & Rogenhofer, 1875) (Cryptolechia) Brazil (Am)

No.	Name	Distribution
	perfusa Meyrick, 1916	French Guiana
580	**frontalis** (Zeller, 1855) (Cryptolechia)	"Mexico"
581	**fulta** Meyrick, 1926	Colombia
582	**fumifica** (Walsingham, 1912) (Stenoma)	Mexico (Ver)
	submersa (Meyrick, 1915) (Stenoma)	Guyana
*583	**furcata** (Walsingham, 1889) (Stenoma)	USA (Az)
*584	**fuscorectangulata** Duckworth, 1964	USA (Az)
585	**generatrix** Meyrick, 1926	Brazil (RS)
586	**glaphyrodes** (Meyrick, 1913) (Stenoma)	Brazil (Am)
587	**glaucescens** (Meyrick, 1916) (Stenoma)	French Guiana
588	**glycerostoma** Meyrick, 1915	Colombia
589	**graphopterella** (Walker, 1864) (Cryptolechia)	Brazil (Am)
	crocuta (R. Felder & Rogenhofer, 1875) (Cryptolechia)	Brazil (Am)
590	**gravescens** Meyrick, 1926	Colombia
591	**griseanomima** Busck, 1934, repl. name	Surinam
	griseana (Sepp, [1852]-55) (Phalaena), preocc. (not Fabricius, 1794)	Surinam
592	**gubernatrix** Meyrick, 1925	Peru
593	**gymnolopha** Meyrick, 1925	Brazil (Am)
594	**gypsoterma** (Meyrick, 1915) (Stenoma)	Guyana
595	**habilis** (Meyrick, 1915) (Stenoma)	Guyana
596	**haesitans** (Walsingham, 1912) (Aedemoses)	Mexico (Dur)
	hessitans (Heinrich, 1921) (Aedemoses), missp.	
	hesitans Busck, 1934, missp.	
597	**haplocentra** Meyrick, 1925	Brazil (Pa)
598	**hapsicora** Meyrick, 1915	Brazil (SP)
599	**helicias** Meyrick, 1916	French Guiana
600	**hemibathra** Meyrick, 1932	"Mexico"
601	**hemiscia** (Walsingham, 1912) (Stenoma)	Guatemala
602	**herilis** R. Felder & Rogenhofer, 1875	Brazil
	basimacula Möschler, 1882	Surinam
	xanthoptila Meyrick, 1912	Surinam
603	**heterosaris** (Meyrick, 1915) (Stenoma)	Guyana
604	**himaea** Meyrick, 1916	French Guiana
605	**homologa** (Meyrick, 1915) (Aphanoxena)	Guyana
606	**horizontias** (Meyrick, 1925) (Stenoma)	Brazil (Am)
607	**humerella** (Walker, 1864) (Cryptolechia)	Brazil (Am)
	intermedia (R. Felder & Rogenhofer, 1875) (Cryptolechia)	"Ceylon" [S. Am.]
	luscina (Zeller, 1877) (Stenoma)	Panama
	meridiana (Meyrick, 1915) (Stenoma)	Guyana
608	**hyalophanta** (Meyrick, 1932) (Stenoma)	Peru
609	**hydrophora** Meyrick, 1925	Peru
610	**ianthina** (Walsingham, 1913) (Stenoma)	Panama
611	**illepida** Meyrick, 1916	French Guiana
	martini Amsel, 1956	Venezuela
612	**imminens** (Meyrick, 1915) (Stenoma)	Surinam
613	**immota** Meyrick, 1916	French Guiana
614	**impactella** (Walker, 1864) (Cryptolechia)	Brazil (Am)
615	**impedita** (Meyrick, 1915) (Stenoma)	Peru
616	**incisurella** (Walker, 1864) (Cryptolechia)	Brazil (Am)
617	**incompleta** Meyrick, 1932	"Mexico"
618	**incongrua** Meyrick, 1932	Peru
619	**incrassata** Meyrick, 1916	French Guiana
620	**indicatella** (Walker, 1864) (Cryptolechia)	Brazil (Am)
621	**infecta** (Meyrick, 1930) (Stenoma)	Brazil (Pa)
622	**infrenata** (Meyrick, 1918) (Stenoma)	French Guiana
623	**innexa** (Meyrick, 1925) (Stenoma)	Peru
624	**inquinula** Zeller, 1854	Brazil (Pa)
625	**insidiata** (Meyrick, 1916) (Stenoma)	French Guiana
626	**insimulata** Meyrick, 1916	Colombia
627	**intersecta** (Meyrick, 1916) (Stenoma)	French Guiana
628	**iopetra** (Meyrick, 1932) (Stenoma)	Guatemala
629	**ioptila** (Meyrick, 1915) (Stenoma)	Guyana
630	**iras** Meyrick, 1926	Peru
*631	**irene** (Barnes & Busck, 1920) (Stenoma)	USA (Tx)
632	**irenias** (Meyrick, 1916) (Stenoma)	French Guiana
633	**isochyta** (Meyrick, 1915) (Stenoma)	Guyana
634	**isomeris** (Meyrick, 1912) (Stenoma)	Brazil (RJ)
635	**isoplintha** (Meyrick, 1925) (Stenoma)	Brazil (Am)
636	**isoporphyra** (Meyrick, 1932) (Asapharcha)	Costa Rica
637	**isosticta** (Meyrick, 1932) (Stenoma)	"Mexico"
638	**isotona** Meyrick, 1932	Panama
639	**ithytona** Meyrick, 1929	Colombia
640	**juvenalis** (Meyrick, 1930) (Stenoma)	Brazil (Pa)
641	**lacera** (Zeller, 1877) (Auxocrossa)	S. Am. (?)
642	**lampyridella** (Busck, 1914) (Stenoma)	Panama
643	**lathiptila** (Meyrick, 1915) (Stenoma)	Guyana
644	**laudata** Meyrick, 1916	French Guiana
645	**laxa** (Meyrick, 1915) (Stenoma)	Venezuela
646	**lebetias** (Meyrick, 1915) (Stenoma)	French Guiana
647	**lecithaula** Meyrick, 1915	Guyana
648	**lepidocarpa** (Meyrick, 1930) (Stenoma)	Brazil (Pa)
649	**leptogramma** (Meyrick, 1916) (Psephomeres)	French Guiana
*650	**leucillana** (Zeller, 1854) (Cryptolechia)	USA (Ga)
	algidella (Walker, 1864) (Cryptolechia)	Canada
651	**leucocryptis** (Meyrick, 1932) (Stenoma)	Colombia
652	**lignicolor** Zeller, 1877	Peru
	emollita Meyrick, 1926	Colombia
*653	**lindseyi** (Barnes & Busck, 1920) (Stenoma)	USA (Az)
654	**lophoptycha** (Meyrick, 1925) (Stenoma)	Brazil (Am)
655	**lophosaris** (Meyrick, 1925) (Stenoma)	Brazil (Pa)
656	**loxogrammos** (Zeller, 1854) (Cryptolechia)	Brazil
657	**lucrosa** (Meyrick, 1925) (Stenoma)	Brazil (Am)
658	**lunimaculata** (Dognin, 1913) (Stenoma)	Colombia
659	**lysimeris** Meyrick, 1915	Peru
660	**machetes** (Walsingham, 1912) (Stenoma)	Mexico (Gue)
661	**macronota** (Meyrick, 1912) (Stenoma)	Colombia
662	**malachita** Meyrick, 1915	Guyana
663	**manceps** Meyrick, 1925	Peru
664	**marmorea** R. Felder & Rogenhofer, 1875	Brazil (Am)
	mentigera Meyrick, 1926	Bolivia
	quatiens Meyrick, 1930	Brazil (Pa)
	thalamobathra Meyrick, 1932	Bolivia
665	**melanarma** Meyrick, 1916	French Guiana
666	**melanopis** Meyrick, 1909	Peru
667	**melinopa** (Meyrick, 1925) (Stenoma)	Brazil (Pa)
668	**mendax** (Zeller, 1855) (Cryptolechia)	Brazil (MG)
	crypsithias (Meyrick, 1930) (Stenoma)	Brazil (MG)
669	**mesosaris** (Meyrick, 1925) (Stenoma)	French Guiana
670	**mesostrota** Meyrick, 1912	Venezuela
671	**microtypa** (Meyrick, 1915) (Stenoma)	Guyana
672	**milictis** Meyrick, 1925	Brazil (Am)
673	**mitratella** (Busck, 1914) (Stenoma)	Panama
674	**modulata** (Meyrick, 1915) (Stenoma)	Guyana
675	**monocolona** Meyrick, 1932	Bolivia
	monoclona Busck, 1934, missp.	
676	**monosaris** (Meyrick, 1915) (Stenoma)	Guyana
677	**mundella** (Walker, 1864) (Cryptolechia)	Brazil (Am)
	contortella (Walker, 1864) (Paepia)	Brazil (Am)
678	**murinella** (Walker, 1864) (Cryptolechia)	Brazil (Am)
679	**mustella** (Walsingham, 1912) (Stenoma)	Panama
680	**navicularis** (Meyrick, 1930) (Stenoma)	Brazil (SP)
681	**neocrossa** (Meyrick, 1925) (Stenoma)	Peru
682	**nephelocyma** (Meyrick, 1930) (Stenoma)	Brazil (SC)
683	**nerteropa** Meyrick, 1915	Peru
684	**neurographa** Meyrick, 1923	Brazil (RJ)
685	**nictitans** (Zeller, 1854) (Mesoptycha)	Brazil (Pa)
	heteropa (Meyrick, 1913) (Stenoma)	Guyana
686	**nimbata** Meyrick, 1925	Peru
687	**nitescens** Meyrick, 1925	Brazil (Pa)
688	**nitidorella** (Walker, 1864) (Cryptolechia)	Brazil (Am)
689	**nitrota** Meyrick, 1916	French Guiana
690	**notogramma** (Meyrick, 1930) (Stenoma)	Brazil (Pa)
691	**notosaris** (Meyrick, 1925) (Stenoma)	Brazil (Pa)
692	**notosemia** (Zeller, 1877) (Stenoma)	Colombia
693	**nuclearis** Meyrick, 1913	Peru
694	**obtusa** (Meyrick, 1916)(Stenoma)	French Guiana
695	**ocellifer** (Walsingham, 1912) (Stenoma)	Mexico (Dur)
696	**ogmolopha** (Meyrick, 1930) (Stenoma)	Brazil (Am)
697	**ogmosaris** (Meyrick, 1915) (Stenoma)	Guyana
698	**ophrysta** Meyrick, 1912	Surinam
699	**orgadopa** (Meyrick, 1925) (Stenoma)	Brazil (Pa)
700	**orthophaea** Meyrick, 1930	Brazil (SP)
701	**orthotona** Meyrick, 1916	French Guiana
702	**orthriopa** Meyrick, 1925	Brazil (Pa)
703	**ostodes** (Walsingham, 1913) (Stenoma)	Guatemala
704	**ovulifera** (Meyrick, 1925) (Stenoma)	Peru
705	**oxycentra** Meyrick, 1916	French Guiana
706	**oxydecta** (Meyrick, 1915) (Stenoma)	Guyana
707	**oxyschista** (Meyrick, 1925) (Stenoma)	Brazil (Am)
708	**pactota** Meyrick, 1915	Guyana
709	**palaestrias** Meyrick, 1916	French Guiana
710	**pallicosta** (R. Felder & Rogenhofer, 1875) (Cryptolechia)	Brazil (Am)
711	**paracrypta** Meyrick, 1915	Guyana
712	**paracta** (Meyrick, 1915) (Stenoma)	Peru
713	**parastis** Gyen, 1913	Chile
714	**particularis** (Zeller, 1877) (Cryptolechia)	Panama
715	**pellocoma** (Meyrick, 1915) (Aphanoxena)	Guyana
716	**percnocarpa** (Meyrick, 1925) (Stenoma)	Brazil (Am)
717	**percnogona** Meyrick, 1925	Peru
718	**periphrictis** (Meyrick, 1915) (Stenoma)	Guyana
719	**phaeoneura** (Meyrick, 1913) (Stenoma)	Guyana
720	**phaeoplintha** (Meyrick, 1915) (Stenoma)	Guyana
721	**phaeosaris** Meyrick, 1915	Guyana
722	**phaselodes** (Meyrick, 1931) (Stenoma)	Brazil (ES)
	phaseolodes (Busck, 1934) (Stenoma), missp.	
723	**phaula** (Walsingham, 1912) (Stenoma)	Guatemala
724	**phollicodes** (Meyrick, 1916) (Stenoma)	French Guiana
725	**phryactis** Meyrick, 1925	Peru
726	**planicoma** (Meyrick, 1925) (Stenoma)	Brazil (Pa)
727	**platydesma** Meyrick, 1915	Guyana
728	**plerotis** Meyrick, 1925	Peru
729	**plesistia** (Meyrick, 1930) (Stenoma)	Brazil (RJ)
	ptilallactis (Meyrick, 1930) (Stenoma)	Brazil (ES)
730	**plumosa** (Busck, 1914) (Stenoma)	Panama
731	**polyglypta** (Meyrick, 1915) (Stenoma)	Guyana
732	**praecisa** Meyrick, 1912	Brazil (RJ)
733	**praerupta** Meyrick, 1915	Guyana
734	**pratifera** (Meyrick, 1925) (Stenoma)	Costa Rica
735	**prosora** (Walsingham, 1912) (Stenoma)	Panama
736	**protosaris** Meyrick, 1915	Guyana
737	**pseudochyta** Meyrick, 1915	Guyana
738	**ptycta** (Walsingham, 1912) (Athleta)	Guatemala
	cenotes (Walsingham, 1912) (Athleta)	Guatemala
	dryotechna (Meyrick, 1915) (Stenoma)	Guyana

739	**pumilis** (Busck, 1914) (Catarata)	Panama
740	**purulenta** Zeller, 1877	Brazil
741	**pyrgota** (Meyrick, 1930) (Stenoma)	Brazil
742	**pyrobathra** (Meyrick, 1931) (Stenoma)	Brazil (Ba)
743	**pythonaea** Meyrick, 1916	French Guiana
744	**quiescens** (Meyrick, 1916) (Stenoma)	French Guiana
745	**radicalis** (Zeller, 1877) (Stenoma)	Panama
746	**radicicola** Meyrick, 1932	Peru
747	**reciprocella** (Walker, 1864) (Cryptolechia)	Brazil (Pa)
748	**reductella** (Walker, 1864) (Cryptolechia)	Brazil (Am)
	phoebe (Busck, 1911) (Stenoma)	French Guiana
749	**refractrix** Meyrick, 1930	Brazil
750	**renselariana** (Stoll, 1781) (Phalaena)	Surinam
	bahiensis (Perty, [1833]) (Pyralis)	Brazil (Ba)
751	**reprehensa** Meyrick, 1926	Brazil (RS)
	acrobapta Meyrick, 1933	Argentina
752	**resiliens** Meyrick, 1925	Brazil (Pa)
753	**rhipidaula** (Meyrick, 1915) (Stenoma)	Guyana
754	**ribbei** Zeller, 1877	Panama
755	**rostriformis** (Meyrick, 1916) (Stenoma)	French Guiana
756	**sana** Meyrick, 1926	Colombia
757	**sarcinata** Meyrick, 1918	French Guiana
758	**sardania** Meyrick, 1925	Brazil (Pa)
759	**scapularis** (Meyrick, 1918) (Stenoma)	French Guiana
760	**sciospila** (Meyrick, 1930) (Stenoma)	Brazil (ES)
761	**segmentata** (Meyrick, 1915) (Stenoma)	Guyana
762	**sellifera** Meyrick, 1925	Brazil (Am)
763	**semicinerea** Zeller, 1877	Panama
	hemitephras Meyrick, 1930	Brazil (Pa)
764	**semiovata** Meyrick, 1926	Colombia
765	**semisignella** (Walker, 1864) (Cryptolechia)	Brazil (Am)
	consociella (Walker, 1864) (Cryptolechia), preocc. (not Walker, 1864)	Brazil (Am)
	batesella (Walker, 1866) (Cryptolechia), repl. name	Brazil (Am)
	consonella (Busck, 1934) (Stenoma), repl. name	Brazil (Am)
	semisiquella (Busck, 1934) (Stenoma), missp.	
766	**serangodes** Meyrick, 1915	Panama
767	**serarcha** Meyrick, 1930	Brazil (MG)
768	**similis** (Busck, 1911) (Stenoma)	French Guiana
769	**smileuta** Meyrick, 1915	Guyana
770	**sortifera** Meyrick, 1930	Bolivia
771	**sparganota** Meyrick, 1915	Guyana
772	**spermolitha** (Meyrick, 1915) (Stenoma)	Guyana
773	**spurca** (Zeller, 1855) (Cryptolechia)	Venezuela
	humeriferella (Walker, 1864) (Cryptolechia)	Brazil (Am)
	erschoffii (Zeller, 1877) (Stenoma)	Colombia
774	**spurcatella** (Walker, 1864) (Cryptolechia)	Brazil?
	chloromis (Meyrick, 1915) (Stenoma)	Guyana
775	**staurota** Meyrick, 1916	French Guiana
776	**sterrhomitra** (Meyrick, 1925) (Stenoma)	Brazil (Am)
777	**stigmatias** (Walsingham, 1913) (Stenoma)	Guatemala
778	**stringens** Meyrick, 1925	Brazil (Am)
779	**stygeropa** (Meyrick, 1925) (Stenoma)	Brazil (Am)
780	**subdulcis** (Meyrick, 1925) (Stenoma)	Brazil (Pa)
	remorsa (Meyrick, 1925) (Stenoma)	Bolivia
781	**substricta** Meyrick, 1918	French Guiana
782	**suffumigata** Walsingham, 1897	Grenada
783	**superciliosa** Meyrick, 1918	French Guiana
784	**synercta** Meyrick, 1925	Brazil (Pa)
785	**tectoria** (Meyrick, 1915) (Stenoma)	Guyana
786	**teleosema** Meyrick, 1925	Brazil (Pa)
787	**tempestiva** (Meyrick, 1916) (Stenoma)	French Guiana
788	**tephrodesma** (Meyrick, 1916) (Stenoma)	French Guiana
789	**tetrapetra** (Meyrick, 1925) (Stenoma)	Brazil (Am)
790	**thammii** Zeller, 1877	Peru
791	**thapsinopa** Meyrick, 1916	French Guiana
	clivosa Meyrick, 1918	French Guiana
792	**theoretica** Meyrick, 1932	Panama
793	**thesmophora** Meyrick, 1915	Guyana
794	**thylacosaris** (Meyrick, 1915) (Stenoma)	Guyana
795	**thysanodes** (Meyrick, 1915) (Stenoma)	Guyana
	cyanopis (Meyrick, 1915) (Stenoma)	Guyana
796	**tibialis** Zeller, 1877	Brazil
797	**tinactis** (Meyrick, 1915) (Stenoma)	Guyana
798	**tornogramma** Meyrick, 1925	Brazil (Pa)
799	**tractrix** Meyrick, 1925	Brazil (Pa)
800	**tremulella** (Walker, 1864) (Cryptolechia)	Brazil (Am)
	chelobathra Meyrick, 1916	French Guiana
801	**tribomias** (Meyrick, 1915) (Stenoma)	Guyana
802	**tricapsis** (Meyrick, 1930) (Stenoma)	Brazil (SP)
803	**trichonota** Meyrick, 1926	Brazil (RS)
804	**triplectra** (Meyrick, 1915) (Stenoma)	Guyana
805	**triplintha** (Meyrick, 1916) (Aphanoxena)	French Guiana
806	**tripustulella** (Walker, 1864) (Cryptolechia)	Brazil (Am)
807	**trisecta** (Walsingham, 1912) (Athleta)	Mexico (Tab)
808	**trisinuata** Meyrick, 1930	Brazil
	raricilia Meyrick, 1930	Brazil (MG)
809	**tritogramma** Meyrick, 1925	Brazil (Am)
810	**trivallata** Meyrick, 1934	Costa Rica
811	**trochoscia** Meyrick, 1915	Guyana
812	**tumens** (Meyrick, 1916) (Stenoma)	French Guiana
813	**umbratella** Walker, 1864	Brazil (Am)
	pallulella (Busck, 1914) (Stenoma)	Panama
	lacertosa Meyrick, 1915	Guyana
814	**umbriferella** (Walker, 1864) (Cryptolechia)	Brazil (Am)
815	**unisecta** (Meyrick, 1930) (Stenoma)	Brazil (ES)
816	**vacata** Meyrick, 1925	Grenada
817	**vannifera** (Meyrick, 1915) (Stenoma)	Peru
	asphalopis (Meyrick, 1925) (Stenoma)	French Guiana
818	**venatum** (Busck, 1911) (Stenoma)	French Guiana
819	**venezuelensis** Amsel, 1956	Venezuela
820	**virens** (Meyrick, 1912) (Prasolithites)	Colombia
	viridis (Busck, 1914) (Stenoma)	Panama
821	**walchiana** (Stoll, 1782) (Phalaena)	Surinam
	dorsella (Fabricius, 1787) (Tinea)	French Guiana
	griseana (Fabricius, 1794) (Pyralis)	"West Indies"
	dorsella (Fabricius, 1794) (Tinea), redesc.	French Guiana
	lativittella (Walker, 1864) (Cryptolechia)	Brazil
	suppressella (Walker, 1864) (Cryptolechia)	Brazil (Am)
	glaciata Meyrick, 1909	Bolivia
	carphitis Meyrick, 1912	Brazil (SC)
	dynastis Meyrick, 1915	Peru
	ampherista Meyrick, 1925	Surinam
	forsteri Amsel, 1956	Venezuela
822	**xanthopetala** (Meyrick, 1931) (Stenoma)	Brazil
823	**xuthosaris** Meyrick, 1925	Brazil (Am)
824	**xylocosma** Meyrick, 1916	French Guiana
825	**xylurga** (Meyrick, 1913) (Stenoma)	Peru
826	**zanclogramma** (Meyrick, 1915) (Stenoma)	Guyana
827	**zelleri** Walsingham & Durrant, 1896	Panama
	fumipennis (Busck, 1914) (Stenoma)	Panama
828	**zelotes** (Walsingham, 1912) (Stenoma)	Mexico (Gue)

BAEONOMA Meyrick, 1916

829	**euphanes** Meyrick, 1916	French Guiana
830	**favillata** (Meyrick, 1915) (Stenoma)	Peru
831	**helotypa** Meyrick, 1916	French Guiana
832	**holarga** Meyrick, 1916	French Guiana
833	**infamis** Meyrick, 1925	Brazil (Am)
834	**leucodelta** (Meyrick, 1914) (Machimia)	Guyana
835	**leucophaeella** (Walker, 1864) (Cryptolechia)	Brazil (Pa)
836	**mastodes** Meyrick, 1916	French Guiana
837	**orthozona** Meyrick, 1916	French Guiana
838	**suavis** Meyrick, 1916	French Guiana

CATARATA Walsingham, 1912

839	**lepisma** Walsingham, 1912	Panama
840	**obnubila** Busck, 1914	Panama
841	**stenota** Walsingham, 1912	Guatemala

CERCONOTA Meyrick, 1915
Pomphocrita Meyrick, 1930
Cerconata Busck, 1934, missp.

842	**acajuti** Becker, 1971	Brazil (Pr)
843	**achatina** (Zeller, 1855) (Cryptolechia)	Colombia
	lembifera (Meyrick, 1915) (Stenoma)	Guyana
	punicea (Meyrick, 1916) (Stenoma)	French Guiana
844	**agraria** (Meyrick, 1925) (Stenoma)	Bolivia
845	**anonella** (Sepp, [1852]-55) (Phalaena)	Surinam
	hamon (Busck, 1911) (Stenoma)	French Guiana
	strophalodes (Meyrick, 1915) (Stenoma)	Peru
846	**aphanes** (Walsingham, 1912) (Stenoma)	Panama
847	**armiferella** (Walker, 1864) (Cryptolechia)	Brazil (Am)
848	**atricassis** (Meyrick, 1916) (Stenoma)	French Guiana
849	**bathyphaea** (Meyrick, 1932) (Stenoma)	Panama
850	**brachyplaca** (Meyrick, 1926) (Ptilogenes)	Brazil (Am)
851	**capnosphaera** (Meyrick, 1916) (Stenoma)	French Guiana
852	**carbonifer** (Busck, 1914) (Stenoma)	Panama
853	**censoria** (Meyrick, 1915) (Stenoma)	Guyana
854	**certiorata** (Meyrick, 1932) (Stenoma)	Brazil (SC)
855	**congressella** (Walker, 1864) (Cryptolechia)	Brazil (Am)
	cycloptila (Meyrick, 1915) (Stenoma)	Guyana
	tyroxesta (Meyrick, 1925) (Stenoma)	Brazil (Am)
	omphacopa (Meyrick, 1931) (Stenoma)	Bolivia
856	**consobrina** (Meyrick, 1915) (Stenoma)	Guyana
857	**dimorpha** Duckworth, 1962	Ecuador
858	**dryoscia** (Meyrick, 1932) (Stenoma)	"Mexico"
859	**ebenocista** (Meyrick, 1928) (Ptilogenes)	French Guiana
860	**emma** (Busck, 1911) (Gonioterma)	French Guiana
	physotricha (Meyrick, 1915) (Stenoma)	Venezuela
861	**eriacma** (Meyrick, 1915) (Stenoma)	Guyana
862	**fermentata** (Meyrick, 1916) (Stenoma)	French Guiana
863	**figularis** (Meyrick, 1918) (Stenoma)	French Guiana
864	**flexibilis** (Meyrick, 1916) (Stenoma)	French Guiana
865	**fulminata** (Meyrick, 1916) (Stenoma)	French Guiana
866	**fusigera** (Meyrick, 1915) (Stenoma)	French Guiana
867	**hexascia** (Meyrick, 1925) (Stenoma)	Brazil (Am)
868	**horometra** (Meyrick, 1925) (Stenoma)	Peru
869	**hydrelaeas** (Meyrick, 1931) (Stenoma)	French Guiana
870	**impressella** (Walker, 1864) (Cryptolechia)	Brazil (Pa)
	prasoleuca (Meyrick, 1916) (Agriophara)	French Guiana
871	**inturbatella** (Walker, 1864) (Cryptolechia)	Brazil (Am)
	xanthobyrsa (Meyrick, 1915) (Stenoma)	Guyana

872	**ischnoscia** (Meyrick, 1932) (Stenoma)	Brazil (SC)
873	**languescens** (Meyrick, 1915) (Stenoma)	Guyana
874	**lutulenta** (Zeller, 1877) (Cryptolechia)	Brazil
875	**lysalges** (Walsingham, 1913) (Gonioterma)	Panama
876	**machinatrix** (Meyrick, 1925) (Stenoma)	Colombia
877	**melema** (Walsingham, 1913) (Gonioterma)	Panama
	cora (Busck, 1914) (Gonioterma)	Panama
878	**minna** (Busck, 1914) (Gonioterma)	Panama
	orthridia (Meyrick, 1916) (Stenoma)	French Guiana
879	**miseta** (Walsingham, 1913) (Stenoma)	Costa Rica
880	**myrodora** (Meyrick, 1925) (Stenoma)	Brazil (Am)
881	**nimbosa** (Zeller, 1877) (Cryptolechia)	Peru
	vanis (Busck, 1911) (Stenoma)	French Guiana
	bythochroa (Meyrick, 1915) (Gonioterma)	Guyana
882	**nitens** (Butler, 1877) (Cryptolechia)	Brazil (Am)
883	**noverca** (Meyrick, 1916) (Agriophara)	French Guiana
884	**nymphas** (Meyrick, 1916) (Stenoma)	French Guiana
885	**obsordescens** (Meyrick, 1930) (Pomphocrita)	Brazil (Pa)
886	**oceanitis** (Meyrick, 1916) (Stenoma)	French Guiana
887	**palliata** (Walsingham, 1913) (Stenoma)	Guatemala
888	**phaeophanes** (Meyrick, 1912) (Stenoma)	Colombia
889	**ptilosema** Meyrick, 1918	French Guiana
890	**recurrens** (Meyrick, 1925) (Stenoma)	Bolivia
891	**recurvella** (Walker, 1864) (Cryptolechia)	Brazil (Am)
	pseudacma (Meyrick, 1918) (Stenoma)	French Guiana
892	**robiginosa** (Meyrick, 1925) (Stenoma)	Brazil (Am)
893	**rosacea** (Butler, 1877) (Cryptolechia)	Brazil (Am)
	erotarcha (Meyrick, 1915) (Stenoma)	French Guiana
894	**sciaphilina** (Zeller, 1877) (Stenoma)	Panama
	torophragma (Meyrick, 1915) (Stenoma)	Guyana
895	**scolopacina** (Walsingham, 1913) (Stenoma)	Panama
896	**seducta** (Meyrick, 1918) (Stenoma)	French Guiana
897	**siraphora** (Meyrick, 1915) (Stenoma)	French Guiana
	leucosaris (Meyrick, 1925) (Stenoma)	Peru
898	**sphragidopis** (Meyrick, 1915) (Stenoma)	Guyana
899	**stylonota** (Meyrick, 1915) (Stenoma)	Guyana
900	**tabida** (Butler, 1877) (Cryptolechia)	Brazil (Am)
	salutaris (Butler, 1877) (Cryptolechia)	Brazil (Am)
	maroni (Busck, 1911) (Stenoma)	French Guiana
	astacopis (Meyrick, 1930) (Stenoma)	Brazil (Pa)
901	**tholodes** (Meyrick, 1915) (Stenoma)	Guyana
902	**tinctipennis** (Butler, 1877) (Cryptolechia)	Brazil (Am)
903	**tricharacta** (Meyrick, 1925) (Stenoma)	Brazil (Am)
904	**trichoneura** (Meyrick, 1913) (Stenoma)	Venezuela
905	**tridesma** Meyrick, 1915	Guyana
906	**trizeucta** (Meyrick, 1930) (Ptilogenes)	Brazil (Pa)
907	**trochistis** (Meyrick, 1916) (Stenoma)	French Guiana
908	**trymalopa** (Meyrick, 1925) (Stenoma)	Colombia
909	**tumulata** (Meyrick, 1916) (Stenoma)	French Guiana

CHLAMYDASTIS Meyrick, 1916
Ptilogenes Meyrick, 1917

910	**acronitis** (Busck, 1911) (Stenoma)	French Guiana
911	**anamochla** (Meyrick, 1929) (Ptilogenes)	Colombia
912	**ancalota** (Meyrick, 1916) (Agriophara)	French Guiana
913	**apoclina** (Meyrick, 1929) (Ptilogenes)	Colombia
914	**arenaria** (Walsingham, 1913) (Stenoma)	Brazil (ES)
	vividella (Busck, 1914) (Stenoma)	Panama
915	**argocymba** (Meyrick, 1926) (Ptilogenes)	Brazil (Pa)
916	**batrachopis** (Meyrick, 1913) (Agriophara)	Peru
917	**bifida** (Meyrick, 1916) (Agriophara)	French Guiana
918	**byssophanes** (Meyrick, 1926) (Ptilogenes)	Brazil (Am)
919	**caecata** (Meyrick, 1916) (Agriophara)	French Guiana
920	**chionoptila** (Meyrick, 1926) (Ptilogenes)	Brazil (Am)
921	**chionosphena** (Meyrick, 1931) (Ptilogenes)	French Guiana
922	**chlorosticta** (Meyrick, 1913) (Agriophara)	Peru
923	**complexa** (Meyrick, 1916) (Agriophara)	French Guiana
924	**crateroptila** (Meyrick, 1918) (Ptilogenes)	French Guiana
925	**curviliniella** (Busck, 1914) (Catarata)	Panama
926	**cystiodes** (Meyrick, 1916) (Agriophara)	French Guiana
927	**deflexa** (Meyrick, 1916) (Agriophara)	French Guiana
928	**deflua** (Meyrick, 1918) (Ptilogenes)	French Guiana
929	**discors** (Meyrick, 1913) (Agriophara)	Peru
930	**disticha** (Meyrick, 1916) (Agriophara)	French Guiana
931	**dominicae** Duckworth, 1969	Dominica
932	**dryosphaera** (Meyrick, 1926) (Ptilogenes)	Brazil (Am)
933	**elaeostola** (Meyrick, 1930) (Ptilogenes)	Brazil (Pa)
934	**epophrysta** (Meyrick, 1909) (Stenoma)	Peru
935	**forcipata** (Meyrick, 1913) (Agriophara)	Colombia
936	**fragmentella** (Dognin, 1913) (Stenoma)	French Guiana
	ponderata (Meyrick, 1916) (Agriophara)	French Guiana
937	**funicularis** (Meyrick, 1926) (Ptilogenes)	French Guiana
938	**galeomorpha** (Meyrick, 1931) (Ptilogenes)	Brazil (Pe)
939	**gemina** (Zeller, 1855) (Cryptolechia)	Colombia
940	**hemichlora** (Meyrick, 1916) (Agriophara)	French Guiana
941	**ichthyodes** (Meyrick, 1926) (Ptilogenes)	Peru
942	**illita** (Meyrick, 1926) (Ptilogenes)	Peru
943	**inscitum** (Busck, 1911) (Stenoma)	French Guiana
944	**inspectrix** (Meyrick, 1916) (Agriophara)	French Guiana
945	**lactis** (Busck, 1911) (Stenoma)	French Guiana
946	**leptobelisca** (Meyrick, 1929) (Ptilogenes)	Colombia
947	**leucoplasta** (Meyrick, 1926) (Ptilogenes)	Brazil (Am)
948	**leucoptila** (Meyrick, 1918) (Ptilogenes)	French Guiana
	laetifica (Busck, 1920) (Stenoma)	Costa Rica
949	**lichenias** (Meyrick, 1916) (Agriophara)	French Guiana
950	**lithograpta** (Meyrick, 1913) (Agriophara)	Peru
951	**melanometra** (Meyrick, 1926) (Ptilogenes)	Colombia
952	**melanonca** (Meyrick, 1915) (Agriophara)	Guyana
953	**mendoron** (Busck, 1911) (Stenoma)	French Guiana
954	**metacymba** (Meyrick, 1916) (Agriophara)	French Guiana
955	**metacystis** (Meyrick, 1918) (Ptilogenes)	French Guiana
956	**metamochla** (Meyrick, 1931) (Ptilogenes)	Brazil
957	**mochlopa** (Meyrick, 1915) (Agriophara)	Guyana
958	**molinella** (Stoll, 1781) (Phalaena)	Surinam
	apicalis (Busck, 1911) (Stenoma)	French Guiana
959	**monastra** (Meyrick, 1909) (Stenoma)	Peru
960	**morbida** (Zeller, 1877) (Cryptolechia)	Peru
961	**mysticopis** (Meyrick, 1926) (Ptilogenes)	Peru
962	**nestes** (Busck, 1911) (Stenoma)	French Guiana
963	**obnupta** (Meyrick, 1916) (Agriophara)	French Guiana
964	**ommatopa** (Meyrick, 1926) (Ptilogenes)	Bolivia
965	**ophiopa** (Meyrick, 1916) (Agriophara)	French Guiana
966	**orion** (Busck, 1920) (Stenoma)	Guatemala
	rufispinis (Meyrick, 1932) (Ptilogenes)	Colombia
967	**oxyplaca** (Meyrick, 1929) (Ptilogenes)	Colombia
968	**paradromis** (Meyrick, 1915) (Agriophara)	Colombia
969	**perducta** (Meyrick, 1916) (Agriophara)	French Guiana
970	**phytoptera** (Busck, 1914) (Stenoma)	Panama
971	**platyspora** (Meyrick, 1932) (Ptilogenes)	Brazil (SP)
	amblystoma (Meyrick, 1936) (Ptilogenes)	Brazil (RS)
972	**plocogramma** (Meyrick, 1915) (Agriophara)	Guyana
973	**poliopa** (Meyrick, 1916) (Agriophara)	Colombia
974	**praenubila** (Meyrick, 1926) (Ptilogenes)	Brazil (Am)
975	**prudentula** (Meyrick, 1926) (Ptilogenes)	Peru
976	**ptilopa** (Meyrick, 1913) (Agriophara)	Colombia
977	**rhomaeopa** (Meyrick, 1931) (Ptilogenes)	Brazil (ES)
978	**scutellata** (Meyrick, 1916) (Agriophara)	French Guiana
979	**smodicopa** (Meyrick, 1915) (Agriophara)	Brazil (Am)
980	**spectrophthalma** (Meyrick, 1932) (Stenoma)	Bolivia
981	**squamosa** (Walsingham, 1892) (Diastoma)	St. Vincent
982	**stagnicolor** (Meyrick, 1926) (Ptilogenes)	Brazil (Am)
983	**steloglypta** (Meyrick, 1931) (Ptilogenes)	French Guiana
984	**strabonia** (Meyrick, 1930) (Ptilogenes)	Brazil (Pa)
985	**synedra** (Meyrick, 1916) (Agriophara)	Paraguay
986	**trastices** (Busck, 1911) (Stenoma)	French Guiana
	aphrogenes (Meyrick, 1915) (Agriophara)	Guyana
987	**tritypa** (Meyrick, 1909) (Stenoma)	Peru
988	**truncatula** (Meyrick, 1913) (Agriophara)	Venezuela
989	**tryphon** (Busck, 1920) (Stenoma)	Guatemala
990	**ungulifera** (Meyrick, 1929) (Ptilogenes)	Colombia
991	**xylinaspis** (Meyrick, 1915) (Agriophara)	Peru

DIASTOMA Möschler, 1882

992	**nubilella** Möschler, 1882	Surinam

ENERGIA Walsingham, 1912

993	**inopina** Walsingham, 1912	Panama
994	**subversa** Walsingham, 1912	Mexico (Ver)

FALCULINA Zeller, 1877

995	**antitypa** Meyrick, 1917	French Guiana
996	**bella** Duckworth, 1966	Brazil (Am)
997	**caustopis** Meyrick, 1932	Brazil (Am)
998	**kasyi** Duckworth, 1966	Surinam
999	**lepidota** Meyrick, 1916	French Guiana
1000	**ochricostata** Zeller, 1877	C. Am. (?)

GONIOTERMA Walsingham, 1897

1001	**advocata** (Meyrick, 1916) (Stenoma)	French Guiana
1002	**aesiocopia** (Walsingham, 1913) (Stenoma)	Mexico (Ver)
	aphrogramma (Meyrick, 1929) (Stenoma)	Panama
1003	**alsiosum** Walsingham, 1913	Panama
1004	**anna** Busck, 1911	Surinam
1005	**argiceraunа** (Meyrick, 1925) (Stenoma)	Colombia
1006	**bolistis** (Meyrick, 1925) (Stenoma)	Brazil (Am)
1007	**bryophanes** (Meyrick, 1915) (Stenoma)	French Guiana
1008	**burmanniana** (Stoll, 1781) (Phalaena)	Surinam
	tortricella (Walker, 1864) (Cryptolechia)	Brazil (Am)
1009	**chlorina** (Kearfott, 1911) (Stenoma)	Brazil (SP)
1010	**choleroptila** (Meyrick, 1915) (Stenoma)	Guyana
1011	**chromolitha** (Meyrick, 1925) (Stenoma)	Bolivia
1012	**compressa** (Walsingham, 1913) (Stenoma)	Mexico (Tab)
	cacoeciella (Amsel, 1956) (Stenoma)	Venezuela
1013	**conchita** Busck, 1920	Guatemala
	desidiosa (Meyrick, 1925) (Stenoma)	Colombia
1014	**crocoptila** (Meyrick, 1915) (Stenoma)	Guyana
1015	**descitum** Walsingham, 1913	Panama
1016	**diatriba** (Walsingham, 1913) (Stenoma)	Guatemala
1017	**dimetropis** (Meyrick, 1932) (Stenoma)	Mexico (Gue)
1018	**expansa** (Meyrick, 1915) (Stenoma)	Brazil (RJ)
1019	**exquisita** Duckworth, 1964	Brazil (SP)
1020	**fastigata** (Meyrick, 1915) (Stenoma)	Guyana
1021	**gubernata** (Meyrick, 1915) (Stenoma)	French Guiana

1022 **ignobilis** (Zeller, 1854) (Cryptolechia) Brazil
pauperatella (Walker, 1864) (Cryptolechia) Brazil (Am)
anita Busck, 1920 Guatemala
1023 **indecora** (Zeller, 1854) (Cryptolechia) Brazil
1024 **inga** Busck, 1911 French Guiana
1025 **latipennis** (Zeller, 1877) (Cryptolechia) Colombia
algosa (Meyrick, 1916) (Stenoma) French Guiana
1026 **linteata** (Meyrick, 1916) (Stenoma) French Guiana
1027 **notifera** (Meyrick, 1915) (Stenoma) Paraguay
1028 **pacatum** Walsingham, 1913 Guatemala
1029 **periscelta** (Meyrick, 1915) (Stenoma) Peru
1030 **phortax** Meyrick, 1915 Guyana
ochrosaris (Meyrick, 1925) (Stenoma) Brazil (Am)
1031 **pleonastes** (Meyrick, 1915) (Stenoma) French Guiana
1032 **projecta** (Meyrick, 1915) (Stenoma) French Guiana
1033 **seppiana** (Stoll, 1781) (Phalaena) Surinam
platycolpa (Meyrick, 1915) (Stenoma) French Guiana
sceptrifera (Meyrick, 1916) (Stenoma) French Guiana

HYALOPSEUSTIS Meyrick, 1925

1034 **vitrea** Meyrick, 1925 Peru

LETHATA Duckworth, 1964

1035 **aletha** Duckworth, 1967 Peru
1036 **amazona** Duckworth, 1967 Brazil (Am)
1037 **angusta** Duckworth, 1967 Brazil (Pr)
1038 **anophthalma** (Meyrick, 1931) (Stenoma) Paraguay
badiella(Amsel, 1956) (Stenoma) Venezuela
maculata Duckworth, 1964 Brazil (SC)
1039 **aromatica** (Meyrick, 1915) (Stenoma) Brazil (SP)
1040 **asthenopa** (Meyrick, 1916) (Stenoma) French Guiana
1041 **bovinella** (Busck, 1914) (Stenoma) Panama
curiata (Meyrick, 1929) (Stenoma) Panama
indistincta (Amsel, 1956) (Stenoma) Venezuela
1042 **buscki** Duckworth, 1964 Belize
1043 **dispersa** Duckworth, 1967 Brazil (MT)
1044 **fernandezyepezi** Duckworth, 1967 Venezuela
1045 **fusca** Duckworth, 1964 Brazil (Pa)
1046 **glaucopa** (Meyrick, 1912) (Stenoma) Colombia
1047 **gypsolitha** (Meyrick, 1931) (Stenoma) Paraguay
1048 **herbacea** (Meyrick, 1931) (Stenoma) Brazil (SP)
1049 **illustra** Duckworth, 1967 Peru
1050 **irresoluta** Duckworth, 1967 Peru
1051 **lanosa** Duckworth, 1967 Panama
1052 **leucothea** (Busck, 1914) (Stenoma) Panama
1053 **monopa** Duckworth, 1967 Brazil (RJ)
1054 **mucida** Duckworth, 1967 Colombia
1055 **myopina** (Zeller, 1877) (Cryptolechia) Brazil
1056 **myrochroa** (Meyrick, 1915) (Stenoma) Venezuela
1057 **obscura** Duckworth, 1967 Peru
1058 **oculosa** Duckworth, 1967 Brazil (SP)
1059 **optima** Duckworth, 1967 Peru
1060 **psidii** (Sepp, [1852]-55) (Phalaena) Surinam
invigilans (Meyrick, 1915) (Stenoma) French Guiana
1061 **pyrenodes** (Meyrick, 1915) (Stenoma) Argentina
1062 **ruba** Duckworth, 1964 Brazil (SC)
1063 **satyropa** (Meyrick, 1915) (Stenoma) French Guiana
1064 **sciophthalma** (Meyrick, 1931) (Stenoma) Brazil
1065 **striolata** (Meyrick, 1932) (Stenoma) Brazil (SC)
1066 **trochalosticta** (Walsingham, 1913) (Stenoma) Panama
trochilosticta (Busck, 1934) (Stenoma), missp.

LOXOTOMA Zeller, 1854

1067 **elegans** Zeller, 1854 Colombia
rhodanthes Meyrick, 1915 Guyana
1068 **seminigrescens** Meyrick, 1932 Brazil (RJ)

MENESTA Clemens, 1860

1069 **astronoma** (Meyrick, 1909) (Stenoma) Bolivia
1070 **cinereocervina** (Walsingham, 1892) (Gelechia) St. Vincent
1071 **succinctella** (Walker, 1864) (Gelechia) Brazil (Am)

MOTHONICA Walsingham, 1912

1072 **cubana** Duckworth, 1969 Cuba
1073 **fluminata** (Meyrick, 1912) (Stenoma) Colombia
1074 **ocellea** Forbes, 1930 Guatemala
1075 **periapta** Walsingham, 1912 Costa Rica

MYSAROMIMA Meyrick, 1926

1076 **liquescens** Meyrick, 1926 Colombia

ORPHNOLECHIA Meyrick, 1909

1077 **acridula** (Meyrick, 1918) (Stenoma) French Guiana
1078 **anaphanta** (Meyrick, 1925) (Stenoma) Brazil (Pa)
1079 **crypsiphragma** Meyrick, 1909 Bolivia
1080 **neastra** (Meyrick, 1915) (Stenoma) Guyana

PARASCAEAS Meyrick, 1936

1081 **uranophanes** (Meyrick, 1931) (Stenoma) Colombia
cyanolampra Meyrick, 1936 Panama

PARASPASTIS Meyrick, 1915

1082 **circographa** Meyrick, 1915 Guyana

PETALOTHYRSA Meyrick, 1921

1083 **microphthalma** Meyrick, 1921 Brazil (Am)

PETASANTHES Meyrick, 1925

1084 **leucactis** Meyrick, 1925 Ecuador

PHELOTROPA Meyrick, 1915

1085 **conversa** Meyrick, 1923 French Guiana
1986 **oenodes** Meyrick, 1915 Guyana

PROMENESTA Busck, 1914

1087 **autampyx** Meyrick, 1925 Peru
citroscia Meyrick, 1931 Brazil (SP)
1088 **callichlora** Meyrick, 1915 Guyana
1089 **capnocoma** (Meyrick, 1931) (Stenoma) Brazil (ES)
1090 **chrysampyx** Meyrick, 1915 Guyana
1091 **haplodoxa** Meyrick, 1925 Brazil (Pa)
1092 **isotrocha** Meyrick, 1918 Argentina
1093 **leucomias** Meyrick, 1925 Brazil (Pa)
1094 **lithochroma** Busck, 1914 Panama
1095 **marginella** Busck, 1914 Panama
1096 **solella** (Walker, 1864) (Gelechia) Brazil (Am)
1097 **triacmopa** (Meyrick, 1931) (Stenoma) Paraguay

RECTIOSTOMA Becker, 1982, repl. name [23]
Setiostoma Zeller, 1875, preocc. (R. Felder & Rogenhofer, 1875 [Glyphipterigidae])

1098 **argyrobasis** (Duckworth, 1971) (Setiostoma) Venezuela
1099 **callidora** (Meyrick, 1909) (Setiostoma) Bolivia
1100 **chrysibasis** (Duckworth, 1971) (Setiostoma) Brazil (SC)
1101 **cirrhobasis** (Duckworth, 1971) (Setiostoma) El Salvador
1102 **cnecobasis** (Duckworth, 1971) (Setiostoma) Bolivia
1103 **earobasis** (Duckworth, 1971) (Setiostoma) Bolivia
1104 **eusema** (Walsingham, 1914) (Setiostoma) Guatemala
1105 **fernaldella** (Riley, 1889) (Setiostoma) USA (Ca)
1106 **flinti** (Duckworth, 1971) (Setiostoma) Mexico (SLP)
1107 **haemitheia** (R. Felder & Rogenhofer, 1875) (Setiostoma) Colombia
chlorobasis (Zeller, 1875) (Setiostoma) [Peru]
haemotheia (Meyrick, 1913) (Setiostoma), missp.
dietzi (Duckworth, 1971) (Setiostoma) Colombia
1108 **leuconympha** (Meyrick, 1921) (Setiostoma) Brazil (Am)
1109 **ochrobasis** (Duckworth, 1971) (Setiostoma) Bolivia
1110 **silvibasis** (Duckworth, 1971) (Setiostoma) Venezuela
1111 **thiobasis** (Duckworth, 1971) (Setiostoma) Brazil (SC)
1112 **xanthobasis** (Zeller, 1875) (Setiostoma) USA (Tx)
1113 **xuthobasis** (Duckworth, 1971) (Setiostoma) Colombia

RHODANASSA Meyrick, 1915

1114 **io** (Busck, 1911) (Stenoma) French Guiana
callimnestra Meyrick, 1915 French Guiana

STENOMA Zeller, 1839 [24]
Auxocrossa Zeller, 1854
Auxocrassa Walsingham, 1882, missp.

1115 **acontiella** (Walker, 1864) (Cryptolechia) Brazil (Am)
1116 **acratodes** Meyrick, 1916 French Guiana
1117 **adminiculata** Meyrick, 1915 French Guiana
1118 **adoratrix** Meyrick, 1925 Bolivia
1119 **adulans** Meyrick, 1925 Peru
malacoxesta Meyrick, 1930 Brazil (Pa)
1120 **adustella** (Walker, 1864) (Cryptolechia) Venezuela
1121 **adytodes** Meyrick, 1925 Peru
1122 **affirmatella** Busck, 1914 Panama
1123 **aggregata** Meyrick, 1916 French Guiana
1124 **albida** (Walker, 1864) (Cryptolechia) Brazil (Am)
1125 **alligans** (Butler, 1877) (Cryptolechia) Brazil (Am)
obelodes Meyrick, 1915 Guyana
1126 **alluvialis** Meyrick, 1925 Peru
1127 **ambiens** Meyrick, 1923 French Guiana
1128 **amphiptera** Meyrick, 1913 Peru
1129 **anaxesta** Meyrick, 1915 Guyana
1130 **ancillaris** Meyrick, 1916 French Guiana
1131 **anconitis** Meyrick, 1915 Guyana
aconitis Clarke, 1955, missp.
1132 **ancylacma** Meyrick, 1925 Peru
1133 **anetodes** Meyrick, 1915 Guyana
1134 **annosa** (Butler, 1877) (Cryptolechia) Brazil (Am)
cirrhogramma Meyrick, 1930 Brazil (Ba)

sublunaris Meyrick, 1930 — Brazil (Pa)
agathelpis Meyrick, 1932 — Brazil (SC)
1135 **antitacta** Meyrick, 1925 — Peru
1136 **aphrophanes** Meyrick, 1918 — Brazil (SP)
1137 **aplytopis** Meyrick, 1930 — Brazil (Pa)
1138 **apsorrhoa** Meyrick, 1915 — Guyana
1139 **aptila** Meyrick, 1915 — Guyana
1140 **argillacea** (Zeller, 1877) (Cryptolechia) — Peru
1141 **argospora** Meyrick, 1915 — Guyana
1142 **armata** (Zeller, 1877) (Cryptolechia) — Brazil (RJ)
1143 **ascodes** Meyrick, 1915 — Guyana
1144 **assignata** Meyrick, 1918 — French Guiana
1145 **aterpes** Walsingham, 1913 — Mexico (Ver)
1146 **augescens** Meyrick, 1925 — Peru
1147 **auricoma** Meyrick, 1930 — Brazil (ES)
1148 **aztecana** Walsingham, 1913 — Mexico (Gue)
1149 **balanoptis** Meyrick, 1932 — Peru
1150 **baliandra** Meyrick, 1915 — Guyana
theobromae (Busck, 1920) (Zetesima) — Surinam
1151 **bathrocentra** Meyrick, 1915 — Guyana
1152 **bathrogramma** (Meyrick, 1912) (Orphnolechia) — Venezuela
1153 **bathyntis** Meyrick, 1931 — Brazil (ES)
1154 **benigna** Meyrick, 1916 — French Guiana
1155 **biannulata** Meyrick, 1930 — Brazil (Pa)
1156 **bicensa** Meyrick, 1915 — Brazil (SP)
1157 **biseriata** (Zeller, 1877) (Cryptolechia) — Brazil
1158 **bisignata** Meyrick, 1916 — French Guiana
1159 **blandula** Meyrick, 1915 — Venezuela
1160 **bryocosma** Meyrick, 1916 — French Guiana
1161 **bryoxyla** Meyrick, 1915 — Peru
1162 **byssina** (Zeller, 1855) (Cryptolechia) — Brazil (Pa)
tetragonella (Walker, 1864) (Cryptolechia) — Brazil (Am)
isabella (R. Felder & Rogenhofer, 1875) (Cryptolechia) — Brazil (Am)
1163 **bythitis** Meyrick, 1915 — Guyana
1164 **caesarea** Meyrick, 1915 — Guyana
1165 **caesia** Meyrick, 1915 — Guyana
1166 **callicoma** Meyrick, 1916 — French Guiana
1167 **camarodes** Meyrick, 1915 — French Guiana
1168 **camptospila** Meyrick, 1925 — Brazil (Am)
1169 **cana** (R. Felder & Rogenhofer, 1875) (Cryptolechia) — Brazil (Am)
octacentra Meyrick, 1915 — Peru
antilyra Meyrick, 1925 — French Guiana
1170 **canonias** Meyrick, 1913 — French Guiana
1171 **capnobola** Meyrick, 1913 — Guyana
1172 **caryodesma** Meyrick, 1925 — Peru
1173 **cassigera** Meyrick, 1915 — French Guiana
1174 **castellana** Meyrick, 1916 — French Guiana
1175 **catenifer** Walsingham, 1912 — Guatemala
1176 **catharmosta** Meyrick, 1915 — Guyana
1177 **cathosiota** Meyrick, 1925 — Brazil (Am)
1178 **chalepa** Walsingham, 1913 — Panama
1179 **chalybaeella** (Walker, 1864) (Cryptolechia) — Brazil (Am)
1180 **charitarcha** Meyrick, 1915 — Guyana
1181 **chionogramma** (Meyrick, 1909) (Orphnolechia) — Bolivia
rectificata Meyrick, 1925 — Peru
1182 **chloroloba** Meyrick, 1915 — Peru
1183 **chloroplaca** (Meyrick, 1915) (Gonioterma) — Guyana
1184 **chloroxantha** Meyrick, 1925 — Brazil (RS)
1185 **cholerocrossa** Meyrick, 1930 — Brazil (ES)
1186 **chromatopa** Meyrick, 1930 — Brazil
1187 **chromotechna** Meyrick, 1925 — Brazil (Am)
1188 **citroxantha** Meyrick, 1916 — French Guiana
1189 **claripennis** Busck, 1914 — Panama
1190 **clysmographa** Meyrick, 1925 — Peru
1191 **codicata** Meyrick, 1916 — French Guiana
1192 **colligata** Meyrick, 1915 — Guyana
1193 **collybista** (Meyrick, 1915) (Gonioterma) — Peru
1194 **columbaris** Meyrick, 1909 — Peru
1195 **comma** Busck, 1911 — French Guiana
melanocrypta Meyrick, 1915 — Guyana
1196 **commutata** (Meyrick, 1926) (Eumiturga) — Brazil (Am)
1197 **completella** (Walker, 1864) (Cryptolechia) — Brazil (Am)
1198 **compsocharis** Meyrick, 1925 — Bolivia
1199 **compsocoma** Meyrick, 1930 — Brazil (Pa)
1200 **condemnatrix** Meyrick, 1930 — Colombia
1201 **congrua** Meyrick, 1925 — Peru
1202 **coniophaea** Meyrick, 1930 — Brazil (Pa)
1203 **consociella** (Walker, 1864) (Cryptolechia) — Brazil (Am)
petrina Walsingham, 1912 — Guatemala
1204 **conveniens** Meyrick, 1925 — Brazil (Am)
1205 **convexicostata** (Zeller, 1877) (Cryptolechia) — Brazil (RJ)
liniella Busck, 1910 — Costa Rica
cantatrix Meyrick, 1925 — Bolivia
1206 **corvula** (Meyrick, 1912) (Antaeotricha) — Colombia
1207 **crambina** Busck, 1920 — Mexico (Gue)
1208 **crepitans** Meyrick, 1918 — French Guiana
crepitana Busck, 1934, missp.
1209 **crocosticta** Meyrick, 1925 — Peru
1210 **crypsangela** Meyrick, 1932 — Peru
1211 **crypsastra** Meyrick, 1915 — Guyana
1212 **crypsetaera** Meyrick, 1925 — Brazil (Am)
1213 **curtipennis** (Butler, 1877) (Cryptolechia) — Brazil (Am)
1214 **cyanarcha** Meyrick, 1915 — Guyana
1215 **cycnographa** Meyrick, 1930 — Brazil (Pa)
1216 **cymbalista** Meyrick, 1918 — French Guiana
1217 **cyphoxantha** Meyrick, 1931 — Brazil (ES)
1218 **dasyneura** Meyrick, 1923 — French Guiana
1219 **decora** (Zeller, 1854) (Cryptolechia) — Brazil
1220 **delphinodes** Meyrick, 1925 — Peru
1221 **deltomis** Meyrick, 1925 — Brazil (Am)
1222 **deuteropa** Meyrick, 1931 — Paraguay
1223 **diametrica** Meyrick, 1926 — Colombia
1224 **dilinopa** Meyrick, 1925 — Brazil (Am)
1225 **diorista** (Meyrick, 1929) (Ptilogenes) — Colombia
1226 **discrepans** Meyrick, 1925 — Peru
discrepana Busck, 1934, missp.
1227 **dispilella** (Walker, 1866) (Cryptolechia) — [Brazil]
1228 **dorcadopa** Meyrick, 1916 — French Guiana
1229 **dryaula** Meyrick, 1925 — Brazil (Pa)
1230 **dryoconis** Meyrick, 1930 — Bolivia
1231 **dryocosma** Meyrick, 1918 — French Guiana
1232 **elaeurga** Meyrick, 1926 — Bolivia
1233 **embythia** Meyrick, 1916 — French Guiana
1234 **eminens** Meyrick, 1918 — French Guiana
1235 **emphatica** Meyrick, 1916 — French Guiana
1236 **empyrota** Meyrick, 1915, repl. name — Argentina
tortricella (Staudinger, 1899) (Cryptolechia), preocc. (not Walker, 1864) — Argentina
1237 **enumerata** Meyrick, 1932 — Brazil (RS)
1238 **epicnesta** Meyrick, 1915 — Guyana
1239 **epicta** Walsingham, 1912 — Mexico (Tab)
1240 **epipacta** Meyrick, 1915 — Guyana
1241 **eumenodora** Meyrick, 1937 — Argentina
1242 **eusticta** Meyrick, 1916 — French Guiana
1243 **eva** Meyrick, 1915 — Guyana
1244 **evanescens** (Butler, 1877) (Cryptolechia) — Brazil (Am)
1245 **exarata** (Zeller, 1854) (Cryptolechia) — Brazil (Pa)
1246 **exempta** Meyrick, 1925 — Brazil
1247 **exhalata** Meyrick, 1915 — French Guiana
1248 **explicita** Meyrick, 1930 — Brazil (Pa)
1249 **externella** (Walker, 1864) (Cryptolechia) — Brazil (Am)
megaspilella (Walker, 1864) (Cryptolechia) — Brazil (Am)
aggerata Meyrick, 1915 — Guyana
1250 **fallax** (Butler, 1877) (Cryptolechia) — Brazil (Am)
1251 **farraria** Meyrick, 1915 — Brazil (SP)
1252 **fassliana** (Dognin, 1913) (Hilarographa) — Colombia
1253 **fenestra** Busck, 1914 — Panama
lithoxesta Meyrick, 1915 — Guyana
1254 **ferculata** Meyrick, 1923 — French Guiana
1255 **ferrocanella** (Walker, 1864) (Cryptolechia) — Brazil (Am)
marcida (Butler, 1877) (Cryptolechia) — Brazil (Am)
ferricanella Walsingham, 1913, emend.
1256 **finitrix** Meyrick, 1925 — Colombia
1257 **frondifer** Busck, 1914 — Panama
1258 **fulcrata** Meyrick, 1915 — French Guiana
1259 **funerana** (Sepp, [1832-40]) (Phalaena) — Surinam
1260 **fusistrigella** (Walker, 1864) (Cryptolechia) — Brazil (Am)
faecosa (R. Felder & Rogenhofer, 1875) (Cryptolechia) — Brazil [Am?]
1261 **futura** Meyrick, 1913 — Peru
1262 **gemellata** Meyrick, 1916 — French Guiana
1263 **grandaeva** (Zeller, 1854) (Cryptolechia) — Brazil (Pa)
chrysogastra Meyrick, 1915 — French Guiana
1264 **graphica** Busck, 1920 — Costa Rica
1265 **gymnastis** Meyrick, 1915 — Guyana
1266 **halmas** Meyrick, 1925 — Peru
1267 **haploxyla** Meyrick, 1915 — Guyana
1268 **harpoceros** Meyrick, 1930 — Brazil (Pa)
1269 **hectorea** (Meyrick, 1915) (Gonioterma) — Peru
1270 **hemilampra** Meyrick, 1915 — French Guiana
1271 **hemiphanta** Meyrick, 1925 — Brazil (Am)
1272 **herifuga** Meyrick, 1932 — Peru
1273 **hesmarcha** (Meyrick, 1930) (Ptilogenes) — Brazil (Pa)
1274 **heteroxantha** Meyrick, 1931 — Paraguay
1275 **himerodes** Meyrick, 1916 — French Guiana
1276 **holcadica** Meyrick, 1916 — French Guiana
1277 **holophaea** (Meyrick, 1916) (Baeonoma) — French Guiana
1278 **homala** Walsingham, 1912 — Mexico
1279 **hopfferi** (Zeller, 1854) (Auxocrossa) — Brazil (Pa)
phyllocosma Meyrick, 1916 — French Guiana
1280 **hoplitica** Meyrick, 1925 — Brazil (Am)
1281 **horocharis** Meyrick, 1930 — Brazil (Pa)
1282 **horocyma** Meyrick, 1925 — Brazil (Am)
1283 **hospitalis** Meyrick, 1915 — Brazil (RJ)
1284 **hyacinthitis** Meyrick, 1930 — Guyana
1285 **hyalocryptis** Meyrick, 1930 — Brazil (Pa)
1286 **hydraena** Meyrick, 1916 — French Guiana
1287 **hypocirrha** Meyrick, 1930 — Brazil (Am)
1288 **iatma** Meyrick, 1915 — French Guiana
1289 **icteropis** Meyrick, 1925 — Brazil (Pa)
1290 **immersa** Walsingham, 1913 — Mexico (Gue)
1291 **immunda** (Zeller, 1854) (Cryptolechia) — Brazil (RJ)
tectella (Walker, 1864) (Cryptolechia) — Brazil (Am)
perophora Meyrick, 1915 — Peru
1292 **impressella** (Busck, 1914) (Gonioterma) — Panama
cecropia Meyrick, 1916, repl. name — Panama
1293 **impurata** Meyrick, 1915 — Guyana

No.	Taxon	Distribution
1294	**inardescens** Meyrick, 1925	Brazil (Am)
1295	**inflata** (Butler, 1877) (Cryptolechia)	Brazil (Am)
	stella (Busck, 1911) (Gonioterma)	French Guiana
	delenita (Butler, 1877) (Cryptolechia)	French Guiana
1296	**infusa** Meyrick, 1916	French Guiana
1297	**injucunda** Meyrick, 1925	Peru
1298	**invulgata** Meyrick, 1915	Venezuela
1299	**iocoma** Meyrick, 1915	French Guiana
1300	**iopercna** Meyrick, 1932	Peru
1301	**iostalacta** Meyrick, 1925	Peru
1302	**irascens** Meyrick, 1930	Brazil (Pa)
1303	**jucunda** Meyrick, 1915	Peru
1304	**klemaniana** (Stoll, 1782) (Phalaena)	Surinam
1305	**lapidea** Meyrick, 1916	French Guiana
1306	**lapilella** (Busck, 1914) (Catarata)	Panama
	involucralis Meyrick, 1931	Brazil (ES)
1307	**latitans** (Dognin, 1905) (Cryptolechia)	Brazil
1308	**lavata** Walsingham, 1913	Mexico (Tab)
1309	**leptogma** Meyrick, 1925	Colombia
1310	**leucana** (Sepp, [1832-40]) (Phalaena)	Surinam
	tenera (Zeller, 1854) (Cryptolechia)	Brazil (Pa)
	virginalis (Butler, 1877) (Cryptolechia)	Brazil (Am)
	neanica Walsingham, 1913	Panama
1311	**leucaniella** (Walker, 1864) (Cryptolechia)	Venezuela
1312	**libertina** Meyrick, 1916	French Guiana
1313	**litura** Zeller, 1839	Brazil
1314	**lucidiorella** (Walker, 1864) (Cryptolechia)	Brazil (Pa)
	cretifera (R. Felder & Rogenhofer, 1875) (Cryptolechia)	Colombia
	javarica (Butler, 1877) (Cryptolechia)	Brazil (Am)
	paramochla Meyrick, 1918	French Guiana
1315	**luctifica** Zeller, 1877	Panama
1316	**macraulax** Meyrick, 1930	Brazil (RJ)
1317	**macroptycha** Meyrick, 1930	Panama
	spermidias Meyrick, 1932	Panama
1318	**melanesia** Meyrick, 1912	Colombia
1319	**melanixa** Meyrick, 1912	Colombia
	acrosticta Walsingham, 1913	Guatemala
1320	**meligrapta** Meyrick, 1925	Brazil
1321	**melixesta** Meyrick, 1925	Colombia
1322	**meridogramma** Meyrick, 1930	Brazil (Pa)
1323	**methystica** Meyrick, 1930	Brazil (Pa)
1324	**metroleuca** Meyrick, 1930	Brazil (Pa)
1325	**meyeriana** (Stoll, 1781) (Phalaena)	Surinam
1326	**milichodes** Meyrick, 1915	Colombia
1327	**minor** Busck, 1914	Panama
1328	**mniodora** Meyrick, 1925	Colombia
1329	**mendula** Meyrick, 1916	French Guiana
1330	**muscula** Zeller, 1877	Panama
	sciocnesta Meyrick, 1925	Brazil (Am)
1331	**myrrhinopa** Meyrick, 1932	Brazil (Am)
1332	**nebrita** Walsingham, 1913	Panama
1333	**negotiosa** Meyrick, 1925	Brazil (Am)
1334	**neopercna** Meyrick, 1930	Brazil (Pa)
1335	**neoptila** Meyrick, 1925	Brazil (Pa)
1336	**neurocentra** Meyrick, 1925	Peru
1337	**neurotona** (Meyrick, 1915) (Athleta)	Guyana
1338	**nigricans** (Busck, 1914) (Athleta)	Panama
1339	**niphacma** Meyrick, 1916	French Guiana
1340	**niphochlaena** (Meyrick, 1926) (Ptilogenes)	Peru
1341	**nycteropa** Meyrick, 1915	Guyana
1342	**oblita** (Butler, 1877) (Cryptolechia)	Brazil (Am)
	patula Meyrick, 1916	French Guiana
1343	**obovata** Meyrick, 1931	Brazil (RS)
1344	**ochlodes** Walsingham, 1912	Panama
1345	**ochricollis** Zeller, 1877	Panama
	marginata Busck, 1914	Panama
	atmodes Meyrick, 1915	Peru
	thymiota Meyrick, 1915	Guyana
1346	**ochropa** Walsingham, 1913	Panama
	ocellata (Busck, 1914) (Catarata)	Panama
1347	**ochrothicta** Meyrick, 1925	Brazil (Pa)
	ochrothicata Busck, 1934, missp.	
1348	**orneopis** Meyrick, 1925	Brazil (Am)
	arridens Meyrick, 1931	French Guiana
1349	**orthocapna** Meyrick, 1912	Guyana
1350	**orthographa** Meyrick, 1925	Brazil (Am)
1351	**ortholampra** Meyrick, 1930	Brazil (SP)
1352	**ovatella** (Walker, 1864) (Cryptolechia)	Brazil (Am)
	genetta (R. Felder & Rogenhofer, 1875) (Cryptolechia)	Brazil (Am)
1353	**oxyscia** Meyrick, 1923	French Guiana
1354	**pantogenes** Meyrick, 1930	Brazil (SC)
1355	**paracapna** Meyrick, 1915	Guyana
1356	**paraplecta** Meyrick, 1925	Brazil (Am)
1357	**pardalodes** Meyrick, 1918	French Guiana
1358	**paropta** Meyrick, 1916	French Guiana
1359	**patens** Meyrick, 1913	Peru
1360	**peccans** (Butler, 1877) (Cryptolechia)	Brazil (Pa)
	binodis Meyrick, 1915	French Guiana
1361	**periaula** Meyrick, 1916	French Guiana
1362	**peridesma** Meyrick, 1925	Brazil (Pa)
1363	**perirrhoa** Meyrick, 1930	French Guiana
1364	**perjecta** Meyrick, 1931	Brazil
1365	**peronia** Busck, 1913	Guyana
	ebria Meyrick, 1915	Venezuela
1366	**persita** Meyrick, 1915	Peru
1367	**pertinax** Meyrick, 1915	Peru
1368	**phaeomystis** Meyrick, 1925	Peru
1369	**phalacropa** Meyrick, 1932	Panama
1370	**phylloxantha** Meyrick, 1933	Argentina
1371	**picrantis** Meyrick, 1930	Brazil (Pa)
1372	**picta** (Zeller, 1854) (Cryptolechia)	Brazil (SP)
1373	**plagosa** (Zeller, 1877) (Antaeotricha)	Brazil
1374	**platyphylla** Meyrick, 1916	French Guiana
1375	**platyterma** Meyrick, 1915	Guyana
1376	**plurima** Walsingham, 1912	Guatemala
1377	**porphyrastis** Meyrick, 1915	Guyana
1378	**praecauta** Meyrick, 1916	French Guiana
1379	**procritica** Meyrick, 1925	Brazil (Am)
1380	**promotella** Zeller, 1877	Panama
	associata Meyrick, 1925	Peru
1381	**psalmographa** Meyrick, 1931	Brazil (ES)
1382	**psilomorpha** Meyrick, 1915	Peru
1383	**ptychobathra** Meyrick, 1930	Brazil (Pa)
1384	**ptychocentra** Meyrick, 1916	French Guiana
1385	**ptychophthalma** Meyrick, 1930	Bolivia
1386	**pustulatella** (Walker, 1864) (Cryptolechia)	Brazil (Am)
1387	**pyramidea** Walsingham, 1913	Guatemala
1388	**pyrrhias** Meyrick, 1915	Guyana
1389	**pyrrhonota** Meyrick, 1915	Guyana
1390	**receptella** (Walker, 1864) (Cryptolechia)	Brazil (Am)
1391	**recondita** Meyrick, 1915	Guyana
1392	**redintegrata** Meyrick, 1925	Bolivia
1393	**regesta** Meyrick, 1926	Peru
1394	**relata** Meyrick, 1925	Brazil (Am)
1395	**residuella** (Zeller, 1877) (Cryptolechia)	S. Am.
1396	**rhodocolpa** Meyrick, 1916	French Guiana
	orthroptila Meyrick, 1936	Peru
1397	**rhothiodes** Meyrick, 1915	Brazil (SP)
1398	**rosa** (Busck, 1911) (Gonioterma)	French Guiana
1399	**sagax** Busck, 1914	Panama
1400	**salome** Busck, 1911	Brazil (Pr)
	clavifera Meyrick, 1912	Brazil (SP)
1401	**salubris** Meyrick, 1925	Brazil (Am)
1402	**satelles** Meyrick, 1925	Brazil (Am)
1403	**sciogama** Meyrick, 1930	Brazil (Ba)
1404	**scitiorella** (Walker, 1864) (Cryptolechia)	Brazil (Am)
	laeviuscula Zeller, 1877	Colombia
	felix Busck, 1914	Panama
	argotoma Meyrick, 1915	Guyana
1405	**scoriodes** (Meyrick, 1915) (Orphnolechia)	Guyana
	avida Meyrick, 1915	Guyana
1406	**secundata** Meyrick, 1925	Peru
1407	**sematopa** Meyrick, 1915	Guyana
1408	**sequestra** Meyrick, 1918	French Guiana
1409	**sericata** (Butler, 1877) (Cryptolechia)	Brazil (Am)
	eminula Meyrick, 1915	French Guiana
1410	**sesquitertia** (Zeller, 1854) (Cryptolechia)	Brazil
	graphiphorella (Walker, 1864) (Cryptolechia)	[Brazil]
1411	**sexmaculata** (Dognin, 1904) (Cryptolechia)	Ecuador
1412	**simplex** Busck, 1914	Panama
1413	**sommerella** (Zeller, 1877) (Cryptolechia)	Brazil
	nymphotima Meyrick, 1931	Brazil (Ba)
	xylograpta Meyrick, 1931	Brazil (SP)
1414	**spodinopis** Meyrick, 1931	Brazil
1415	**stabilis** (Butler, 1877) (Cryptolechia)	Brazil (Am)
	chionodora Meyrick, 1915	Guyana
	rita (Busck, 1920) (Gonioterma)	Guyana
1416	**staudingerana** (Maassen, 1890) (Tortrix)	Colombia
	contophora Meyrick, 1915	French Guiana
	heterosema Meyrick, 1930	Brazil (Pa)
1417	**stephanodes** Meyrick, 1931	French Guiana
1418	**straminella** (Walker, 1864) (Cryptolechia)	Brazil (Am)
1419	**strenuella** (Walker, 1864) (Cryptolechia)	Brazil (Am)
1420	**striatella** Busck, 1914	Panama
1421	**strigivenata** (Butler, 1877) (Cryptolechia)	Brazil (Am)
	urbana (Butler, 1877) (Cryptolechia)	Brazil (Am)
	entephras Meyrick, 1915	Guyana
	porinodes Meyrick, 1915	Peru
	clarkei Amsel, 1956	Venezuela
1422	**stupefacta** Meyrick, 1916	French Guiana
1423	**subita** Meyrick, 1925	Bolivia
1424	**sublimbata** (Zeller, 1877) (Antaeotricha)	Panama
1425	**subnotatella** (Walker, 1864) (Cryptolechia)	Brazil (Am)
1426	**surinamella** (Möschler, 1882) (Cryptolechia)	Surinam
	expilata Meyrick, 1915	French Guiana
1427	**sustentata** Meyrick, 1926	Colombia
1428	**symmicta** Walsingham, 1913	Panama
1429	**symphonica** Meyrick, 1916	French Guiana
1430	**symposias** (Meyrick, 1915) (Gonioterma)	Colombia
	nepheloleuca Meyrick, 1932	Costa Rica
1431	**syngraphopis** Meyrick, 1930	Brazil (Pa)
1432	**tetrabola** Meyrick, 1913	Peru
1433	**thaleropa** Meyrick, 1916	French Guiana
1434	**thespia** Meyrick, 1915	French Guiana
1435	**thologramma** Meyrick, 1932	Peru
1436	**thoristes** Busck, 1911	French Guiana

No.	Taxon	Distribution
1437	**thylacandra** Meyrick, 1915	French Guiana
1438	**tolmeta** Walsingham, 1912	Panama
1439	**trichocolpa** Meyrick, 1915	French Guiana
	centrodina Meyrick, 1916	French Guiana
1440	**trichorda** Meyrick, 1912	Colombia
1441	**trilineata** (Butler, 1877) (Cryptolechia)	Brazil (Am)
1442	**tripustulata** (Zeller, 1854) (Cryptolechia)	Venezuela
1443	**trirecta** Meyrick, 1931	Brazil
1444	**tristrigata** (Zeller, 1854) (Cryptolechia)	Brazil
	aphanodesma Meyrick, 1915	French Guiana
1445	**tyrocrossa** Meyrick, 1925	Brazil (Am)
1446	**ulosema** Meyrick, 1930	Brazil (RJ)
1447	**umbrinervis** Meyrick, 1930	Brazil (Pa)
1448	**uncticoma** Meyrick, 1916	French Guiana
1449	**unguentata** Meyrick, 1930	Brazil (Pa)
1450	**uruguayensis** (Berg, 1885) (Cryptolechia)	Uruguay
1451	**vaccula** Walsingham, 1913	Mexico (Dur)
1452	**vaga** (Butler, 1877) (Cryptolechia)	Brazil (Am)
	licmaea Meyrick, 1915	Guyana
1453	**vapida** (Butler, 1877) (Cryptolechia)	Brazil (Am)
	acribota Meyrick, 1930	Brazil (Pa)
1454	**vasifera** Meyrick, 1925	Colombia
	unisignis Meyrick, 1932	Bolivia
1455	**ventilatrix** Meyrick, 1916	French Guiana
1456	**vexata** Meyrick, 1915	Guyana
1457	**vinifera** Meyrick, 1916	French Guiana
1458	**viridiceps** (R. Felder & Rogenhofer, 1875) (Anatolmis)	Colombia
	erythropennis (Dognin, 1914) (Dasycera)	Colombia
1459	**vita** (Busck, 1911) (Gonioterma)	French Guiana
1460	**vitreola** Meyrick, 1925	Brazil (Pa)
1461	**vivax** Busck, 1914	Panama
1462	**volitans** Meyrick, 1925	Peru
1463	**xanthophaeella** (Walker, 1864) (Paepia)	Brazil (Am)
1464	**xylinopa** Meyrick, 1925	Bolivia
1465	**ybyrajuba** Becker, 1971	Brazil (MT)
1466	**zephyritis** Meyrick, 1925	Brazil (Pa)
1467	**zobeida** Meyrick, 1931	Mexico (Tab)
	orthopa Meyrick, 1932	Panama

THIOSCELIS Meyrick, 1909

No.	Taxon	Distribution
1468	**directrix** Meyrick, 1909	Peru
1469	**fuscata** Duckworth, 1967	Peru
1470	**geranomorpha** Meyrick, 1932	Brazil (RJ)
1471	**lipara** Duckworth, 1967	Brazil (Am)
1472	**whalleyi** Duckworth, 1967	Brazil (MG)

TIMOCRATICA Meyrick, 1912
Lychnocrates Meyrick, 1926

No.	Taxon	Distribution
1473	**agramma** Becker, 1982	Brazil (ES)
1474	**albella** (Zeller, 1839) (Depressaria)	Surinam
1475	**albitogata** Becker, 1982	Brazil (Pr)
1476	**amseli** Duckworth, 1962, repl. name	Venezuela
	albella Amsel, 1956, preocc. (not Zeller, 1839)	Venezuela
1477	**anelaea** (Meyrick, 1932) (Stenoma)	Brazil (Pa)
1478	**argonais** (Meyrick, 1925) (Stenoma)	Brazil (Am)
	argonias Clarke, 1955, missp.	
1479	**bicornuta** Becker, 1982	Brazil (RJ)
1480	**butyrota** (Meyrick, 1929) (Stenoma)	Colombia
	syndicastis (Meyrick, 1929) (Stenoma)	Colombia
1481	**constrictivalva** Becker, 1982	Ecuador
1482	**effluxa** (Meyrick, 1930) (Lychnocrates)	Bolivia
1483	**fraternella** (Busck, 1910) (Stenoma)	Costa Rica
1484	**fuscipalpalis** Becker, 1982	Venezuela
1485	**grandis** (Perty, [1833]) (Yponomeuta)	Brazil (Pi)
1486	**guarani** Becker, 1982	Argentina
1487	**isarga** (Meyrick, 1925) (Stenoma)	Bolivia
1488	**leucocapna** (Meyrick, 1926) (Lychnocrates)	Colombia
1489	**leucorectis** (Meyrick, 1925) (Stenoma)	Brazil (MG)
1490	**longicilia** Becker, 1982	Colombia
1491	**loxotoma** (Busck, 1909) (Stenoma)	Mexico
1492	**macroleuca** (Meyrick, 1932) (Stenoma)	Bolivia
1493	**major** (Busck, 1911) (Stenoma)	Peru
1494	**maturescens** (Meyrick, 1925) (Stenoma)	French Guiana
1495	**megaleuca** (Meyrick, 1912) (Stenoma)	Colombia
1496	**melanocosta** Becker, 1982	Brazil (DF)
1497	**melanostriga** Becker, 1982	Brazil (SC)
1498	**meridionalis** Becker, 1982	Brazil (Pr)
1499	**monotonia** (Strand, 1911) (Cryptolechia)	Ecuador
	isographa Meyrick, 1912	Venezuela
	claudescens Meyrick, 1925	Peru
	crassa Meyrick, 1925	Brazil (Pa)
1500	**nivea** Becker, 1982	Brazil (DF)
1501	**palpalis** (Zeller, 1877) (Cryptolechia)	Brazil
	auxoleuca (Meyrick, 1925) (Stenoma)	Brazil (MG)
	haywardi Busck, 1939	Argentina
1502	**parvifusca** Becker, 1982	Costa Rica
1503	**parvileuca** Becker, 1982	Brazil (SP)
1504	**philomela** (Meyrick, 1925) (Stenoma)	Peru
1505	**pompeiana** Meyrick, 1925	Peru
1506	**spinignatha** Becker, 1982	Peru
1507	**subovalis** (Meyrick, 1932) (Stenoma)	Brazil (Am)
	stomatocosma (Meyrick, 1932) (Stenoma)	Brazil (Am)
1508	**titanoleuca** Becker, 1982	Peru
1509	**venifurcata** Becker, 1982	Brazil (DF)
1510	**xanthosoma** (Dognin, 1913) (Stenoma)	French Guiana
	a) **xanthosoma** (Dognin, 1913) (Stenoma)	French Guiana
	sacra (Meyrick, 1918) (Stenoma)	French Guiana
	b) **leucocephala** Becker, 1982	Venezuela
1511	**xanthotarsa** Becker, 1982	Panama

ZETESIMA Walsingham, 1912

No.	Taxon	Distribution
1512	**albipes** (R. Felder & Rogenhofer, 1875) (Simaethis)	Brazil (Am)
1513	**lasia** Walsingham, 1912	Panama
	patellifera (Meyrick, 1931) (Stenoma)	Brazil (RS)
1514	**portentosa** Busck, 1914	Panama
1515	**scoliandra** (Meyrick, 1915) (Stenoma)	Guyana

Oecophorinae

Tribe Oecophorini

ACARTOPHILA Meyrick, 1932

No.	Taxon	Distribution
1516	**microsacta** Meyrick, 1932	Costa Rica
1517	**stauromacha** Meyrick, 1932	Colombia

ALICIANA Clarke, 1978

No.	Taxon	Distribution
1518	**albella** (Blanchard, 1852) (Epigraphia)	Chile
1519	**geminata** Clarke, 1978	Chile

ALTIURA Clarke, 1978

No.	Taxon	Distribution
1520	**maculata** Clarke, 1978	Chile

ALYNDA Clarke, 1978

No.	Taxon	Distribution
1521	**cinnamomea** Clarke, 1978	Chile
1522	**sarissa** Clarke, 1978	Chile
1523	**striata** Clarke, 1978	Chile

ANIUTA Clarke, 1978

No.	Taxon	Distribution
1524	**melanoma** Clarke, 1978	Chile
1525	**ochroleuca** Clarke, 1978	Chile

ARCTOPODA Butler, 1883
Polypseustis Dognin, 1908

No.	Taxon	Distribution
1526	**maculosa** Butler, 1883	Chile
	cuprea (Dognin, 1908) (Polypseustis)	Chile

ATELOSTICHA Meyrick, 1883
Crypsynarthra Lower, 1901

No.	Taxon	Distribution
1527	**depressariella** (Walker, 1864) (Cryptolechia)	Brazil (Am)
1528	**ochrospora** Meyrick, 1928	Brazil (Am)
1529	**percnotoxa** Meyrick, 1920	Brazil (Am)

ATHA Clarke, 1978

No.	Taxon	Distribution
1530	**trimacula** Clarke, 1978	Chile

ATOPOTORNA Meyrick, 1932

No.	Taxon	Distribution
1531	**ptychoptila** Meyrick, 1932	Panama

BORKHAUSENIA Hübner, [1825]
Amaurosetia Stephens, [1835]
Pseudatemelia Rebel, 1910

No.	Taxon	Distribution
1532	**bryotrophoides** (Zeller, 1877) (Oecophora)	Colombia
1533	**commixta** Meyrick, 1928	Peru
1534	**confarreatella** (Zeller, 1877) (Oecophora)	Colombia
1535	**crimnodes** Meyrick, 1912	Argentina
1536	**longipalpis** Meyrick, 1931	Argentina
1537	**minnetta** (Butler, 1883) (Oecophora)	Chile
1538	**praesul** Meyrick, 1931	Argentina
1539	**sensilis** Meyrick, 1926	Peru
1540	**syrmeutis** Meyrick, 1931	Argentina

BRYMBLIA Hodges, 1974

No.	Taxon	Distribution
1541	**quadrimaculella** (Chambers, 1875) (Oecophora)	USA (Co)

CALLISTENOMA Butler, 1883

No.	Taxon	Distribution
1542	**ustimacula** (Zeller, 1874) (Cryptolechia)	Chile
	zelleri Butler, 1883	Chile

CECIDOLECHIA Kieffer & Jörgensen, 1910
Cecidolechia Strand, 1911

No.	Taxon	Distribution
1543	**maculicostella** Kieffer & Jörgensen, 1910	Argentina
	maculicostella Strand, 1911	Argentina

CHEZALA Walker, 1864

1544 **lunularis** Meyrick, 1926 — Colombia
lunaris Gaede, 1938, missp.

CORITA Clarke, 1978

1545 **amphichroma** Clarke, 1978 — Chile

DECANTHA Busck, 1914

1546 **minuta** Busck, 1914 — Panama

DEIA Clarke, 1978

1547 **lineola** Clarke, 1978 — Chile

DESPINA Clarke, 1978

1548 **rhodosema** (Meyrick, 1931) (Borkhausenia) — Chile

DINOTROPA Meyrick, 1916

1549 **ochrocrossa** Meyrick, 1916 — French Guiana

DITA Clarke, 1978

1550 **fasciatipedella** (Zeller, 1874) (Cryptolechia) — Chile
1551 **phococara** Clarke, 1978 — Chile

DOLIOTECHNA Meyrick, 1914

1552 **bimarginata** Meyrick, 1929 — Colombia
1553 **designata** Meyrick, 1914 — Guyana
1554 **eucentra** Meyrick, 1920 — Brazil (Pa)
1555 **hyalophaea** Meyrick, 1914 — Guyana
1556 **integra** Meyrick, 1914 — Guyana
1557 **orphnopis** Meyrick, 1914 — Guyana
1558 **spilocrossa** Meyrick, 1920 — Peru
1559 **triplacodes** Meyrick, 1920 — Brazil (Am)
1560 **trissobathra** Meyrick, 1920 — Peru
1561 **virginea** Meyrick, 1914 — Guyana

DYSGNORIMA Zeller, 1877

1562 **subannulata** Zeller, 1877 — Colombia

ENDROSIS Hübner, [1825]

1563 **sarcitrella** (Linnaeus, 1758) (Phalaena) — [Sweden?]
sarcitella (Linnaeus, 1761) (Phalaena), missp.
fenestrella (Scopoli, 1763) (Phalaena) — [Yugoslavia]
lactella ([Denis & Schiffermüller], 1775) (Tinea) — [Austria]
betulinella (Hübner, [1818-19]) (Tinea) — Europe
kennicottella Clemens, 1860 — USA (Il)
subditella (Walker, 1864) (Gelechia) — New Zealand
lacteella [sic] var. antarctica Staudinger, 1899 — Chile

EOMICHLA Meyrick, 1916
Astoxena Meyrick, 1930

1564 **argentidisca** (Dognin, 1905) (Borkhausenia) — Brazil (SP)
1565 **clotho** (Meyrick, 1930) (Astoxena) — French Guiana
1565.1 **hallwachsae** Clarke, 1983 — Costa Rica
1566 **imperiella** (Busck, 1914) (Peleopoda) — Panama
1567 **irenella** (Busck, 1911) (Peleopoda) — French Guiana
1568 **leucoclista** Meyrick, 1930 — Brazil (Pa)
1569 **maroniella** (Busck, 1911) (Peleopoda) — French Guiana
1570 **notandella** (Busck, 1911) (Peleopoda) — French Guiana
1571 **nummulata** Meyrick, 1916 — French Guiana
1572 **regiella** (Busck, 1912) (Peleopoda) — Panama
1573 **thysiarcha** Meyrick, 1928 — Bolivia
1574 **xystidota** Meyrick, 1918 — French Guiana

ERAINA Clarke, 1978

1575 **thamnocephala** Clarke, 1978 — Chile

EXOSPHRANTIS Meyrick, 1931

1576 **bibula** Meyrick, 1931 — Argentina

HALIMARMARA Meyrick, 1931

1577 **atrivallata** Meyrick, 1931 — Brazil (Am)

HARPELLA Schrank, 1802

1578 **semnodoxa** Meyrick, 1931 — Chile

HELIOSTIBES Zeller, 1874

1579 **mathewi** Zeller, 1874 — Chile

HOFMANNOPHILA Spuler, 1910

1580 **pseudospretella** (Stainton, 1849) (Oecophora) — UK (England)

HYPERSKELES Butler, 1883

1581 **choreutidea** Butler, 1883 — Chile

IFEDA Hodges, 1966

1582 **perobtusa** (Meyrick, 1922) (Batrachedra) — Brazil (Am)

INGA Busck, 1908
Doxa Walsingham, 1912
Lysigrapha Meyrick, 1914
Pelomimas Meyrick, 1914
Orsimacha Meyrick, 1914
Siderograptis Meyrick, 1920
Phanerodoxa Meyrick, 1921
Epimoryctis Meyrick, 1930
Horomeristis Meyrick, 1931
Agriotorna Meyrick, 1931

1583 **analis** (Busck, 1914) (Cryptolechia) — Panama
1584 **ancorata** (Walsingham, 1912) (Cryptolechia) — Costa Rica
capsaria (Meyrick, 1914) (Lysigrapha) — Guyana
1585 **brevisella** (Walker, 1864) (Cryptolechia) — Brazil (Am)
1586 **callierastis** (Meyrick, 1920) (Machimia) — Brazil (Am)
1587 **calycocentra** (Meyrick, 1931) (Horomeristis) — Peru
1588 **camelopis** (Meyrick, 1920) (Atelosticha) — Brazil (Am)
1589 **cancanodes** (Meyrick, 1918) (Machimia) — French Guiana
1590 **catasticta** (Meyrick, 1920) (Atelosticha) — Brazil (Pa)
1591 **caumatias** (Meyrick, 1929) (Machimia) — Colombia
1592 **cerophaea** (Meyrick, 1914) (Machimia) — Guyana
1593 **chlorochroa** (Meyrick, 1912) (Machimia) — Argentina
1594 **cnecodes** (Meyrick, 1920) (Atelosticha) — Brazil (Am)
1595 **concinna** (Meyrick, 1912) (Machimia) — Surinam
1596 **conserva** (Meyrick, 1914) (Machimia) — Guyana
1597 **corystes** (Meyrick, 1914) (Machimia) — Guyana
1598 **crossota** (Walsingham, 1912) (Cryptolechia) — Guatemala
1599 **crucifera** (Busck, 1914) (Psilocorsis) — Panama
1600 **cupidinea** (Meyrick, 1914) (Machimia) — Guyana
1601 **custodita** (Meyrick, 1928) (Machimia) — Brazil (Am)
1602 **cyclophthalma** (Meyrick, 1916) (Atelosticha) — French Guiana
cyclophthalama Hodges, 1972, missp.
1603 **deligata** (Meyrick, 1914) (Machimia) — Guyana
1604 **dilecta** (Meyrick, 1920) (Machimia) — Peru
1605 **distorta** (Meyrick, 1920) (Machimia) — Peru
1606 **elaphodes** (Meyrick, 1930) (Machimia) — Bolivia
1607 **empyrea** (Meyrick, 1920) (Machimia) — Brazil (Pa)
1608 **encamina** (Meyrick, 1912) (Machimia) — Venezuela
1609 **entaphrota** (Meyrick, 1915) (Machimia) — Colombia
1610 **erasicosma** (Meyrick, 1916) (Atelosticha) — French Guiana
1611 **eriocnista** (Meyrick, 1931) (Agriotorna) — Brazil (Pa)
1612 **erotias** (Meyrick, 1912) (Machimia) — Colombia
1613 **erythema** (Walsingham, 1912) (Cryptolechia) — Guatemala
marcella (Busck, 1914) (Cryptolechia) — Panama
1614 **fervida** (Zeller, 1855) (Cryptolechia) — Brazil
furvida (Walker, 1864) (Cryptolechia), missp.
1615 **flava** (Zeller, 1839) (Depressaria) — Brazil
1616 **fundigera** (Meyrick, 1912) (Machimia) — Surinam
1617 **furva** (Meyrick, 1916) (Machimia) — French Guiana
1618 **genuina** (Meyrick, 1914) (Machimia) — Guyana
1619 **haemataula** (Meyrick, 1912) (Machimia) — Brazil (SP)
haematula Hodges, 1972, missp.
1620 **halosphora** (Meyrick, 1916) (Atelosticha) — French Guiana
1621 **helobia** (Meyrick, 1931) (Machimia) — Paraguay
1622 **hyperbolica** (Meyrick, 1928) (Machimia) — Brazil (Am)
1623 **icterota** (Meyrick, 1914) (Machimia) — Guyana
1624 **incensatella** (Walker, 1864) (Cryptolechia) — Brazil (Am)
1625 **inflammata** (Meyrick, 1916) (Machimia) — French Guiana
1626 **iracunda** (Meyrick, 1914) (Orsimacha) — Guyana
1627 **lacunata** (Meyrick, 1914) (Machimia) — Guyana
1628 **languida** (Meyrick, 1912) (Machimia), **n. comb.** — Venezuela
1629 **leptophragma** (Meyrick, 1920) (Siderograptis) — Brazil (Pa)
1630 **libidinosa** (Meyrick, 1926) (Machimia) — Colombia
1631 **loxobathra** (Meyrick, 1915) (Cryptolechia) — French Guiana
1632 **meliacta** (Meyrick, 1914) (Machimia) — Guyana
1633 **mercata** (Meyrick, 1914) (Machimia) — Guyana
1634 **mimobathra** (Meyrick, 1920) (Atelosticha) — Brazil (Am)
1635 **mixadelpha** (Meyrick, 1914) (Pelomimas) — Guyana
1636 **molifica** (Meyrick, 1914) (Machimia) — Peru
1637 **molybdopa** (Meyrick, 1920) (Siderograptis) — Brazil (Am)
1638 **mydopis** (Meyrick, 1914) (Machimia) — Guyana
1639 **neospila** (Meyrick, 1928) (Machimia) — Brazil (Am)
1640 **orthodoxa** (Meyrick, 1912) (Machimia) — Brazil (RJ)
1641 **orthophragma** (Meyrick, 1916) (Atelosticha) — French Guiana
1642 **pachybathra** (Meyrick, 1921) (Phanerodoxa) — Brazil (Am)
1643 **pagana** (Meyrick, 1916) (Machimia) — French Guiana
1644 **pagidotis** (Meyrick, 1918) (Machimia) — French Guiana
1645 **percnorma** (Meyrick, 1930) (Epimoryctis) — Brazil (Pe)
1646 **pericyclota** (Meyrick, 1920) (Machimia) — Peru
1647 **perioditis** (Meyrick, 1928) (Machimia) — Peru
1648 **petasodes** (Meyrick, 1914) (Orsimacha) — Guyana

1649 **phaeocrossa** (Meyrick, 1912) (Machimia) Brazil (RJ)
1650 **plectanota** (Meyrick, 1918) (Machimia) Guyana
1651 **porpotis** (Meyrick, 1914) (Machimia) Guyana
1652 **pyrothyris** (Meyrick, 1916) (Machimia) French Guiana
1653 **pyrrhoxantha** (Meyrick, 1931) (Machimia) French Guiana
1654 **refuga** (Meyrick, 1916) (Machimia), **n. comb.** French Guiana
1655 **rhodoclista** (Meyrick, 1916) (Atelosticha) French Guiana
1656 **rosea** (Meyrick, 1920) (Machimia) Brazil (Am)
1657 **roseomarginella** (Busck, 1911) (Cryptolechia) French Guiana
1658 **ruricola** (Meyrick, 1914) (Machimia) Venezuela
1659 **satura** (Meyrick, 1914) (Machimia) Guyana
1660 **sciocrates** (Meyrick, 1929) (Machimia) Colombia
1661 **sciotoxa** (Meyrick, 1914) (Machimia) Guyana
1662 **semotella** (Walker, 1864) (Gelechia) Brazil (Am)
1663 **separatella** (Walker, 1864) (Gelechia) Brazil (Am)
1664 **signifera** (Meyrick, 1914) (Machimia) Guyana
1665 **sodalis** (Walsingham, 1912) (Doxa) Brazil (Pa)
relicta (Meyrick, 1916) (Heliocausta) Guyana
1666 **sparsiciliella** (Clemens, 1864) (Anesychia) USA (Va)
contrariella (Walker, 1864) (Cryptolechia) "?"
inscitella (Walker, 1864) (Cryptolechia) "?"
atropicta (Zeller, 1875) (Cryptolechia) "N. Am."
castigata (Meyrick, 1926) (Cryptolechia) Costa Rica
1667 **speculatrix** (Meyrick, 1914) (Machimia) Guyana
1668 **staphylitis** (Meyrick, 1916) (Machimia) French Guiana
1669 **stativa** (Meyrick, 1920) (Atelosticha) Peru
1670 **stereodesma** (Meyrick, 1916) (Machimia) French Guiana
1671 **taboga** (Busck, 1914) (Epicallima) Panama
1672 **textrina** (Meyrick, 1914) (Machimia) Guyana
1673 **thermoxantha** (Meyrick, 1914) (Lysigrapha) Peru
1674 **trailii** (Butler, 1877) (Cryptolechia) Brazil (Am)
corallina (Meyrick, 1914) (Machimia) Guyana
traili (Meyrick, 1922) (Machimia), missp.
1675 **trifurcata** (Meyrick, 1918) (Machimia) French Guiana
1676 **trygaula** (Meyrick, 1912) (Machimia) Brazil (SP)
1677 **tubicen** (Meyrick, 1921) (Phanerodoxa) Peru
1678 **versatilis** (Meyrick, 1921) (Phanerodoxa) Brazil (Pa)
1679 **virginia** (Busck, 1914) (Doxa) Panama
1680 **voluptaria** (Meyrick, 1914) (Machimia) Guyana

IRENIA Clarke, 1978

1681 **curvula** Clarke, 1978 Chile
1682 **leucoxantha** Clarke, 1978 Chile

LELITA Clarke, 1978

1683 **acmaea** Clarke, 1978 Chile

LYGRONOMA Meyrick, 1913

1684 **cyanastra** (Meyrick, 1909) (Brenthia) Bolivia
1685 **sporimaea** Meyrick, 1913 Guyana

MACAROCOSMA Meyrick, 1931

1686 **philochrysa** Meyrick, 1931 Chile

MATHILDANA Clarke, 1941

1687 **auricollis** (Walsingham, 1912) (Dasycera) [25] Guatemala

MELOCHRYSIS Meyrick, 1916

1688 **heliaca** Meyrick, 1916 French Guiana

ODONNA Clarke, 1982

1689 **passiflorae** Clarke, 1982 Colombia
1690 **xenodora** Clarke, 1982 Colombia

PACHYPHOENIX Butler, 1883
Tyriomorpha Meyrick, 1918
Mattea Duckworth, 1966

1691 **phoenissa** (Butler, 1883) (Cryptolechia) Chile
1692 **sanguinea** Butler, 1883 Chile

PHILOMUSAEA Meyrick, 1931
Philomusea Clarke, 1978, missp.

1693 **brachyxista** Meyrick, 1931 Chile
1694 **craterias** Meyrick, 1931 Argentina
1695 **elissa** Meyrick, 1931 Argentina
1696 **incommoda** Meyrick, 1931 Argentina
1697 **meniscogramma** Clarke, 1978 Chile

PSEUDOECOPHORA Staudinger, 1899

1698 **vitellinella** Staudinger, 1899 Argentina

PYCNOTARSA Meyrick, 1920

1699 **hydrochroa** Meyrick, 1920 Brazil (Pa)

1700 **sulphurea** (Busck, 1914) (Cryptolechia) Panama

PYRAMIDOBELA Braun, 1923
Idioptila Meyrick, 1927

1701 **agyrtodes** (Meyrick, 1927) (Idioptila) USA (Tx)
1702 **compulsa** Meyrick, 1931 Chile
1703 **epibryas** Meyrick, 1931 Brazil (SP)
1704 **ochrolepra** Powell, 1973 Mexico (Chia)
1705 **tetraphyta** Meyrick, 1931 Mexico (Hid)

RETHA Clarke, 1978

1706 **rustica** Clarke, 1978 Chile

REVONDA Clarke, 1978

1707 **eschara** Clarke, 1978 Chile

STASIXENA Meyrick, 1930

1708 **subagrestis** Meyrick, 1930 Brazil (RJ)

STRUTHOSCELIS Meyrick, 1913

1709 **acrobatica** Meyrick, 1913 Peru
1710 **semiotarsa** Meyrick, 1916 French Guiana

TERESITA Clarke, 1978

1711 **diffinis** (R. Felder & Rogenhofer, 1875) (Cryptolechia) Chile
1712 **isaura** Clarke, 1978 Chile

THAUMATOLITA Walsingham, 1912

1713 **hamifera** Walsingham, 1912 Mexico (Mor)
1714 **stemonias** Meyrick, 1920 Peru

THEAMA E. M. Hering, 1958

1715 **argyrophorum** E. M. Hering, 1958 Argentina

UTILIA Clarke, 1978

1716 **falcata** Clarke, 1978 Chile
1717 **florinda** Clarke, 1978 Chile
1718 **luridella** (Zeller, 1874) (Cryptolechia) Chile
1719 **ochracea** (Zeller, 1874) (Cryptolechia) Chile

ZYMRINA Clarke, 1978

1720 **xanthosema** (Meyrick, 1931) (Borkhausenia) Chile

Tribe Stathmopodini

ARAUZONA Walker, [1865]

1721 **basalis** Walker, [1865] Brazil (Am)
1722 **moorei** Busck, 1913 Guyana

SNELLENIA Walsingham, 1889

1723 **flavipennis** (R. Felder & Rogenhofer, 1875) (Eretmocera) S. Am.?
1724 **latipes** (Walker, [1865]) (Tinaegeria) Brazil (Pa)

STATHMOPODA Herrich-Schäffer, 1853
Boocara Butler, 1880
Placostola Meyrick, 1887
Erineda Busck, 1909
Agrioscelis Meyrick, 1913
Kakivoria Nagano, 1916

1725 **antischema** Meyrick, 1922 Peru
1726 **filicula** Clarke, 1978 Colombia

TINAEGERIA Walker, 1856

1727 **croconympha** Meyrick, 1921 Bolivia
1728 **fasciata** Walker, 1856 Brazil (Pa)
1729 **nephelozyga** Meyrick, 1930 "Brazil"
1730 **ochracea** Walker, 1856 Brazil (Pa)
aeneiceps (R. Felder & Rogenhofer, 1875) (Eretmocera) Colombia
1731 **pyromantis** Meyrick, 1921 Peru

Tribe Pleurotini

PLEUROTA Hübner, [1825]
Eupleuris Hübner, [1825]
Holoscolia Zeller, 1839
Protasis Herrich-Schäffer, 1853

1732 **albastrigulella** (Kearfott, 1907) (Dorota [sic]) USA (Ca)

22. ELACHISTIDAE

by V. O. Becker

Elachistinae

DICRANOCTETES Braun, 1918
Donacivola Busck, [1934]

1 **saccharella** (Busck, [1934]) (Donacivola) Cuba

ELACHISTA Treitschke, 1833
Aphelosetia Stephens, 1834
Cycnodia Herrich-Schäffer, 1853
Phigalia Chambers, 1875, preocc. (Duponchel, 1829 [Geometridae])
Atachia Wocke, 1876
Neaera Chambers, 1880, preocc. (Robineau-Desvoidy, 1830 [Diptera])
Hecista Wallengren, 1881
Aphigalia Dyar, 1903
Irenicodes Meyrick, 1919
Euproteodes Viette, 1954

2 **albisquamella** Zeller, 1877 Colombia
3 **luciliella** Zeller, 1877 Colombia
4 **tersectella** Zeller, 1877 Colombia

23. BLASTOBASIDAE

by V. O. Becker

Symmocinae

EUPOLELLA Fletcher, 1940, repl. name [26]
Eupolis Meyrick, 1923, preocc. (Cambridge, 1900 [Arachnida])

1 **stygnota** (Walsingham, 1911) (Glyphidocera) Panama

GLYPHIDOCERA Walsingham, 1892
Harpagandra Meyrick, 1918

2 **abiasta** Meyrick, 1936 Venezuela
3 **audax** Walsingham, 1892 St. Vincent
4 **capraria** Meyrick, 1929 Colombia
5 **carribea** Busck, 1910 Trinidad
caribbea Meyrick, 1925, emend.
6 **catectis** Meyrick, 1923 Ecuador
7 **cerochra** Meyrick, 1929 Colombia
8 **crocogramma** Meyrick, 1923 Brazil
9 **cryphiodes** (Meyrick, 1918) (Harpagandra) Guyana
10 **dominicella** Walsingham, 1897 Dominica
11 **drosophaea** Meyrick, 1929 Colombia
12 **elpista** Walsingham, 1911 Panama
13 **eurrhipis** Meyrick, 1929 Colombia
14 **exsiccata** Meyrick, 1914 Guyana
15 **hypochloa** (Walsingham, 1911) (Dichomeris) Mexico (Son)
16 **illiterata** Meyrick, 1929 Panama
17 **indocilis** Meyrick, 1930 Brazil (Pa)
18 **inurbana** Meyrick, 1914 Guyana
19 **lepidocyma** Meyrick, 1929 Colombia
20 **lophandra** Meyrick, 1929 Brazil (Pa)
21 **melithrepta** Meyrick, 1929 Colombia
22 **notolopha** Meyrick, 1929 Brazil (Pa)
23 **orthoctenis** Meyrick, 1923 Brazil (Pa)
24 **orthotenes** Meyrick, 1929 Colombia
25 **percnoleuca** Meyrick, 1923 Brazil (Am)
26 **perobscura** Walsingham, 1911 Mexico (Tab)
27 **psammolitha** Meyrick, 1923 Brazil
28 **ptilostoma** Meyrick, 1935 Argentina
29 **ptychocryptis** Meyrick, 1929 Colombia
30 **recticostella** Walsingham, 1897 Grenada
31 **reparabilis** Walsingham, 1912 Panama
32 **rhypara** Walsingham, 1911 Mexico (Gue)
33 **sagifera** Meyrick, 1929 Colombia
34 **salinae** Walsingham, 1911 Mexico (Oax)
35 **stenomorpha** Meyrick, 1923 Surinam
36 **thyrsogastra** Meyrick, 1929 Peru
37 **trachyacma** Meyrick, 1931 Paraguay
38 **umbrata** Walsingham, 1911 Guatemala
39 **vestita** Walsingham, 1911 Panama
40 **zophocrossa** Meyrick, 1929 Trinidad

OECIA Walsingham, 1897
Macroceras Staudinger, 1876, preocc. (Semper, 1870 [Mollusca])

41 **oecophila** (Staudinger, 1876) (Macroceras) Italy
maculata Walsingham, 1897 Brazil (Pa)

SCEPTEA Walsingham, 1911
Sceptia McDunnough, 1939, missp.

42 **decedens** Walsingham, 1911 Mexico (Tab)

SYMMOCA Hübner, [1825]
Simoca Weiler, 1877, missp.
Parasymmoca Rebel, 1903
Asarista Meyrick, 1935
Conquassata Gozmány, 1957
Symmoletria Gozmány, 1963

43 **vetusta** Meyrick, 1931 Brazil (Ba)

Blastobasinae

Tribe Blastobasini

AUXIMOBASIS Walsingham, 1892

44 **administra** Meyrick, 1922 Peru
45 **agrestis** Meyrick, 1922 Brazil (Pa)
46 **angusta** Meyrick, 1922 Colombia
47 **brevipalpella** Walsingham, 1897 Grenada
48 **coffeaella** Busck, 1925 Brazil (SP)
49 **constans** Walsingham, 1897 Virgin Is. (St. Thomas)
50 **flavicillata** Walsingham, 1897 Virgin Is. (St. Thomas)
51 **flavida** Meyrick, 1922 Brazil (Pa)
52 **incretata** Meyrick, 1931 Peru
53 **insularis** Walsingham, 1897 Virgin Is. (St. Thomas)
54 **liberatella** (Walker, 1864) (Gelechia) Brazil (Am)
55 **normalis** Meyrick, 1918 Ecuador
56 **obstricta** Meyrick, 1918 Guyana
57 **persimilella** Walsingham, 1892 St. Vincent
58 **prolixa** Meyrick, 1922 Peru
59 **variolata** Walsingham, 1897 Virgin Is. (St. Thomas)

BLASTOBASIS Zeller, 1855
Epistetus Walsingham, 1894
Prostesis Walsingham, 1908

60 **aequivoca** Meyrick, 1922 Guyana
61 **anachasta** Meyrick, 1928 Brazil (Am)
62 **aphilodes** Meyrick, 1918 Colombia
63 **argillacea** Walsingham, 1897 Virgin Is. (St. Thomas)
64 **atmosema** Meyrick, 1930 Brazil (Pa)
65 **atmozona** Meyrick, 1939 Argentina
66 **candidata** Meyrick, 1922 Peru
67 **commendata** Meyrick, 1922 Brazil (Am)
68 **cophodes** Meyrick, 1918 Peru
69 **crotospila** Meyrick, 1926 Galapagos Is.
70 **ergastulella** Zeller, 1877 Colombia
71 **gracilis** Walsingham, 1897 Grenada
72 **grenadensis** Walsingham, 1897 Grenada
73 **guilandinae** Busck, 1900 USA (Fl)
74 **lecaniella** Busck, 1913 Guyana
75 **leucogona** Zeller, 1877 Colombia
76 **leucozyga** Meyrick, 1936 Venezuela
77 **neozona** (Meyrick, 1918) (Exinotis) Guyana
78 **ochrobathra** Meyrick, 1921 Guyana
79 **pacalis** Meyrick, 1922 Brazil (Pa)
80 **subolivacea** Walsingham, 1897 Virgin Is. (St. Thomas)
81 **triangularis** Walsingham, 1897 Virgin Is. (St. Thomas)

HOLCOCERA Clemens, 1863
Catacrypsis Walsingham, 1907
Cynotes Walsingham, 1907
Prosodica Walsingham, 1907

82 **adjutrix** Meyrick, 1918 Guyana
83 **aphanes** (Zeller, 1877) (Blastobasis) Colombia
84 **basiplagata** Walsingham, 1912 Guatemala
85 **chloropeda** Meyrick, 1922 Brazil (Pa)
86 **controversella** (Zeller, 1877) (Blastobasis) Colombia
87 **cylindrota** Meyrick, 1918 Colombia
88 **digesta** Meyrick, 1922 Brazil (Pa)
89 **eusaris** Meyrick, 1922 Peru
90 **hemiteles** Walsingham, 1912 Guatemala
91 **homochromatica** Walsingham, 1912 Mexico (Gue)
92 **limicola** Meyrick, 1918 Ecuador
*93 **nana** Dietz, 1910 USA (Tx)
94 **nephalia** Walsingham, 1907 Guatemala
95 **orthophrontis** Meyrick, 1932 Bolivia
96 **percnoscia** Meyrick, 1932 Brazil (SC)
97 **pinae** Amsel, 1962 Guatemala
98 **proagorella** (Zeller, 1877) (Blastobasis) Colombia
99 **pugionaria** Meyrick, 1918 Guyana
*100 **pusilla** Dietz, 1910 USA (Tx)
101 **supletella** (Zeller, 1877) (Blastobasis) Colombia
102 **sympasta** Meyrick, 1918 Peru
103 **titanica** Walsingham, 1912 Mexico (Oax)

ICONISMA Walsingham, 1897

104 **macrocera** Walsingham, 1897 Virgin Is. (St. Thomas)

METALLOCRATES Meyrick, 1930

105 **transformata** Meyrick, 1930 Brazil (Am)

VALENTINIA Walsingham, 1907

106 **bromeliae** Walsingham, 1912 — Mexico (Ver)
107 **neptes** Walsingham, 1912 — Mexico (Gue)
108 **tarachodes** Walsingham, 1912 — Mexico (Gue)

Tribe Pigritiini

PIGRITIA Clemens, 1860

109 **astuta** Meyrick, 1918 — Colombia
110 **biatomella** Walsingham, 1897 — Virgin Is. (St. Thomas)
111 **mediocris** Walsingham, 1897 — Virgin Is. (St. Thomas)
112 **troctis** Meyrick, 1922 — Barbados

24. COLEOPHORIDAE

by V. O. Becker

Coleophorinae

COLEOPHORA Hübner, [1822]
Eupista Hübner, [1825]
Apista Hübner, [1825]
Haploptilia Hübner, [1825]
Porrectaria Haworth, 1828
Damophila Curtis, 1832
Astyages Stephens, 1834
Metallosetia Stephens, 1834
Casas Wallengren, 1881
Casigneta Wallengren, 1881
Corythangela Meyrick, 1897
Casignetella Strand, 1928
Calaritania Mariani, 1943
Heringiella Börner, 1944
Tolleophora Căpuse, 1971
Ionescumia Căpuse, 1971
Stollia Căpuse, 1971
Razowskia Căpuse, 1971
Orghidania Căpuse, 1971
Frederickoenigia Căpuse, 1971
Suireia Căpuse, 1971
Zagulajevia Căpuse, 1971
Amseliphora Căpuse, 1971
Nemesia Căpuse, 1971
Zangheriphora Căpuse, 1971
Bourgogneja Căpuse, 1971
Aureliania Căpuse, 1971
Bacescuia Căpuse, 1971
Klinzigedia Căpuse, 1971
Vladdelia Căpuse, 1971
Klimeschja Căpuse, 1971
Glaseria Căpuse, 1971
Valvulongia Căpuse, 1971
Falkovitshia Căpuse, 1972
Helopharea Falkovitsh, 1972
Cricotechna Falkovitsh, 1972
Plegmidia Falkovitsh, 1972
Agapalsa Falkovitsh, 1972
Phylloschema Falkovitsh, 1972
Bima Falkovitsh, 1972
Systrophoeca Falkovitsh, 1972
Aporiptura Falkovitsh, 1972
Symphypoda Falkovitsh, 1972
Oedicaula Falkovitsh, 1972
Argyractinia Falkovitsh, 1972
Chnoocera Falkovitsh, 1972
Orthographis Falkovitsh, 1972
Phagolamia Falkovitsh, 1972
Monotemachia Falkovitsh, 1972
Corethropoea Falkovitsh, 1972
Characia Falkovitsh, 1972
Perygra Falkovitsh, 1972
Perygridia Falkovitsh, 1972
Luzulina Falkovitsh, 1972
Carpochena Falkovitsh, 1972
Abaraschia Căpuse, 1973
Amselghia Căpuse, 1973
Ardania Căpuse, 1973
Ascleriductia Căpuse, 1973
Baraschia Căpuse, 1973
Benanderpia Căpuse, 1973
Calcomarginia Căpuse, 1973
Caleophora Căpuse, 1973, missp.
Corothropoea Căpuse, 1973
Cornulivalvulia Căpuse, 1973, missp.
Dumitrescumia Căpuse, 1973
Ecebalia Căpuse, 1973
Globulia Căpuse, 1973
Hamuliella Căpuse, 1973
Helvalbia Căpuse, 1973
Ionnemesia Căpuse, 1973
Kasyfia Căpuse, 1973
Kuznetzovvlia Căpuse, 1973
Latisacculia Căpuse, 1973
Longibacillia Căpuse, 1973
Lucidaesia Căpuse, 1973
Lvaria Căpuse, 1973
Membrania Căpuse, 1973
Metapista Căpuse, 1973
Multicoloria Căpuse, 1973
Neugenvia Căpuse, 1973
Nosyrislia Căpuse, 1973
Ortohgraphis Căpuse, 1973, missp.
Oudejansia Căpuse, 1973
Paravalvulia Căpuse, 1973
Patzakia Căpuse, 1973
Postvinculia Căpuse, 1973
Proglaseria Căpuse, 1973
Quadratia Căpuse, 1973
Rhamnia Căpuse, 1973
Sacculia Căpuse, 1973
Scleriductia Căpuse, 1973
Tollsia Căpuse, 1973
Tuberculia Căpuse, 1973
Ulna Căpuse, 1973
Ductispira Căpuse, 1973
Klimeschjosefia Căpuse, 1975

1 **anisota** Meyrick, 1927 — Bermuda
2 **breyeri** Pastrana, 1963 — Argentina
3 **chiarelliae** Pastrana, 1963 — Argentina
4 **decipiens** Walsingham, 1914 — Mexico (Gue)
5 **exarga** Meyrick, 1917 — Colombia
6 **haywardi** Pastrana, 1963 — Argentina
7 **intexta** Meyrick, 1917 — Peru
8 **lepyropis** Meyrick, 1921 — Brazil (Pa)
9 **pelinopis** Meyrick, 1933 — Mexico (DF)
10 **picticornis** Walsingham, 1897[27] — (West Indies)
11 **pulchricornis** Walsingham, 1897 — Virgin Is. (St. Croix/St. Thomas)
12 **zymotica** Meyrick, 1931 — Paraguay

TOCASTA Busck, 1912

13 **priscella** Busck, 1912 — Panama

Batrachedrinae

AMBLYTENES Meyrick, 1930

14 **lunatica** Meyrick, 1930 — Brazil (Pa)

BATRACHEDRA Herrich-Schäffer, 1853
Eustaintonia Spuler, 1910

15 **calator** Hodges, 1966 — Mexico (Hid)
16 **comosae** Hodges, 1966 — Puerto Rico
*17 **concitata** Meyrick, 1928 — USA (NM)
18 **conspersa** Meyrick, 1916 — Ecuador
19 **copia** Clarke, 1957 — Mexico (Mex)
20 **daduchus** Hodges, 1966 — Jamaica
*21 **decoctor** Hodges, 1966, extralim.
a) **bermudensis** Hodges, 1966 — Bermuda
22 **dolichoscia** Meyrick, 1928 — Colombia
*23 **enormis** Meyrick, 1928 — USA (NM)
24 **knabi** (Walsingham, 1909) (Homaledra) — Mexico (Ver)
25 **linaria** Clarke, 1957 — Mexico (BCS)
26 **meator** Hodges, 1966 — St. Lucia
27 **nuciferae** Hodges, 1966 — Brazil (Ba)
28 **paritor** Hodges, 1966 — Jamaica
29 **repertor** Hodges, 1966 — Panama
30 **rixator** Hodges, 1966 — Colombia
31 **theca** Clarke, 1957 — Mexico (DF)

CHEDRA Hodges, 1966

32 **delector** Hodges, 1966 — Chile

25. MOMPHIDAE

by V. O. Becker

ANCHIMOMPHA Clarke, 1965

1 **melaleuca** Clarke, 1965 — Chile (Juan Fernandez Is.)

MOMPHA Hübner, [1825]
Laverna Curtis, 1839
Lophoptilus Sircom, 1848
Cyphophora Herrich-Schäffer, 1853
Psacaphora Herrich-Schäffer, 1853
Anybia Stainton, 1854
Wilsonia Clemens, 1864
Leucophryne Chambers, 1875

2 **agonistes** (Walsingham, 1909) (Laverna) — Guatemala
3 **basisignella** (Zeller, 1877) (Laverna) — Colombia
4 **conspersa** (Walsingham, 1892) (Anybia) — St. Vincent

5 **constellaris** Meyrick, 1931 Chile
6 **conviva** Meyrick, 1931 Argentina
7 **crassinodis** (Zeller, 1877) (Laverna) Colombia
8 **eloisella** (Clemens, 1860) (Laverna) USA [Pa?]
magnatella (Zeller, 1873) (Phyllocnistis) USA (Ma)
oenotheraeella (Chambers, 1875) (Laverna), repl. name USA (Tx)
lyonetiella (Chambers, 1875) (Laverna) USA (Tx)
9 **exodias** Meyrick, 1931 Argentina
10 **exsultans** (Zeller, 1877) (Laverna) Colombia
11 **farinacea** (Walsingham, 1909) (Laverna) Mexico (Gue)
12 **heterolychna** Meyrick, 1922 Peru
13 **laevinella** (Zeller, 1877) (Laverna) Colombia
14 **leucochrysis** Meyrick, 1935 Argentina
15 **metallifera** (Walsingham, 1897) (Anybia) Jamaica
16 **musota** Meyrick, 1917 Peru
17 **obsessa** (Walsingham, 1909) (Laverna) Mexico (Gue)
18 **ochrosemia** (Zeller, 1877) (Laverna) Colombia
19 **permota** Meyrick, 1917 Colombia
20 **pernota** (Walsingham, 1909) (Laverna) Mexico (Gue)
21 **phalaropis** Meyrick, 1922 Peru
22 **piperatella** (Walsingham, 1897) (Anybia) Virgin Is. (St. Thomas)
23 **polygoni** (Zeller, 1877) (Asychna) Colombia
24 **sectifera** Meyrick, 1922 Brazil (Am)
25 **sympotica** Meyrick, 1931 Chile
26 **tripunctata** (Walsingham, 1897) (Anybia) Virgin Is. (St. Thomas)
27 **trithalama** Meyrick, 1922 Brazil (Am)
28 **verruculella** (Zeller, 1877) (Laverna) Colombia

SYNALLAGMA Busck, 1907

29 **busckiella** Engel, 1907 USA (Pa)
busckellum Walsingham, 1909, emend.

26. AGONOXENIDAE

by V. O. Becker

Blastodacninae

HOMOEOPREPES Walsingham, 1909
Homeoprepes Hodges, 1978, missp.

1 **felisae** Clarke, 1962 Colombia
2 **sympatrica** Clarke, 1962 Colombia
3 **trochiloides** Walsingham, 1909 Costa Rica

MICROCOLONA Meyrick, 1897

4 **transennata** Meyrick, 1922 Brazil (Pa)

NANODACNA Clarke, 1964

5 **ancora** Clarke, 1964 Chile (Juan Fernandez Is.)
6 **indiscriminata** Clarke, 1965 Chile (Juan Fernandez Is.)
7 **logistica** (Meyrick, 1931) (Colonophora) Argentina
8 **vinacea** (Meyrick, 1922) (Homaledra) Peru

NICANTHES Meyrick, 1928

9 **rhodoclea** Meyrick, 1928 Guyana

PAMMECES Zeller, 1863

10 **albivitella** Zeller, 1863 Venezuela
11 **citraula** Meyrick, 1922 Peru
12 **crocoxysta** Meyrick, 1922 Brazil (Am)
13 **lithochroma** Walsingham, 1897 Dom. Rep.
14 **pallida** Walsingham, 1897 Virgin Is. (St. Thomas)
15 **phlogophora** Walsingham, 1909 Panama
16 **problema** Walsingham, 1915 Colombia

PANCLINTIS Meyrick, 1929

17 **socia** Meyrick, 1929 Colombia

PROCHOLA Meyrick, 1915

18 **aedilis** Meyrick, 1915 Guyana
19 **agypsota** Meyrick, 1922 Brazil (Am)
20 **basichlora** Meyrick, 1922 Brazil (Pa)
21 **catacentra** Meyrick, 1922 Brazil (Am)
22 **catholica** Meyrick, 1917 Guyana
23 **chloropis** Meyrick, 1922 Brazil (Am)
24 **euclina** Meyrick, 1922 Peru
25 **fuscula** Forbes, 1931 Puerto Rico
26 **holomorpha** Meyrick, 1931 Paraguay
27 **obstructa** Meyrick, 1915 Ecuador
28 **ochromicta** Meyrick, 1922 Brazil (Am)
29 **oppidana** Meyrick, 1915 Guyana
30 **orphnopa** Meyrick, 1922 Brazil (Am)
31 **orthobasis** Meyrick, 1922 Brazil (Pa)
32 **pervallata** Meyrick, 1922 Brazil (Pa)
33 **prasophanes** Meyrick, 1922 Brazil (Pa)
34 **revecta** Meyrick, 1922 Guyana
35 **semialbata** Meyrick, 1922 Brazil (Am)
36 **sollers** Meyrick, 1917 Guyana
37 **subtincta** (Meyrick, 1922) (Syntetrernis) Brazil (Am)

SYNTETRERNIS Meyrick, 1922

38 **neocompsa** Meyrick, 1933 Argentina
39 **xiphodes** Meyrick, 1922 Peru

ZARATHA Walker, 1864

40 **macrocera** R. Felder & Rogenhofer, 1875 Brazil (Am)
41 **mesonyctia** Meyrick, 1909 Bolivia
42 **pterodactylella** Walker, 1864 Brazil (Am)
niveiventris R. Felder & Rogenhofer, 1875 Colombia

27. COSMOPTERIGIDAE

by V. O. Becker

Antequerinae

ANTEQUERA Clarke, 1941

1 **acertella** (Busck, 1913) (Semioscopis) USA (Ca)
exstimulata (Meyrick, 1928) (Borkhausenia) Mexico (BCN)

EUCLEMENSIA Grote, 1878, repl. name
Hamadryas Clemens, 1864, preocc. (Hübner, 1806 [Nymphalidae])

*2 **schwarziella** Busck, [1901] USA (Az)

Cosmopteriginae

ANONCIA Clarke, 1941

3 **chordostoma** (Meyrick, 1912) (Cryptolechia) Argentina
4 **diveni** (Heinrich, 1921) (Borkhausenia) USA (Tx)

APHANOSARA Forbes, 1931

5 **planistes** Forbes, 1931 Puerto Rico

COSMOPTERIX Hübner, [1825]
Cosmopteryx Zeller, 1839, emend.

6 **abnormalis** Walsingham, 1897 (Cosmopteryx [sic]) Haiti
7 **apiculata** Meyrick, 1922 (Cosmopteryx [sic]) Brazil (Am)
8 **astrapias** Walsingham, 1909 (Cosmopteryx [sic]) Mexico (Tab)
9 **attenuatella** (Walker, 1864) (Gelechia) Jamaica
flavofasciata Wollaston, 1879 (Cosmopteryx [sic]) St. Helena
mimetis Meyrick, 1897 (Cosmopteryx [sic]) Australia
antillia Forbes, 1931 (Cosmopteryx [sic]) Puerto Rico
superba Gozmány, 1960 (Cosmopteryx [sic]) Egypt
10 **callichalca** Meyrick, 1922 (Cosmopteryx [sic]) Brazil (Am)
11 **citrinopa** Meyrick, 1915 (Cosmopteryx [sic]) Peru
12 **diaphora** Walsingham, 1909 (Cosmopteryx [sic]) Mexico (Gue)
13 **erasmia** Meyrick, 1915 (Cosmopteryx [sic]) Guyana
*14 **facunda** Hodges, 1962 USA (Tx)
15 **inaugurata** Meyrick, 1922 (Cosmopteryx [sic]) Brazil (Pa)
16 **interfracta** Meyrick, 1922 (Cosmopteryx [sic]) Brazil (Pa)
17 **irrubricata** Walsingham, 1909 (Cosmopteryx [sic]) Mexico (Ver)
18 **isotoma** Meyrick, 1915 (Cosmopteryx [sic]) Guyana
19 **nyctiphanes** Meyrick, 1915 (Cosmopteryx [sic]) Ecuador
20 **ochleria** Walsingham, 1909 (Cosmopteryx [sic]) Mexico (Tab)
21 **pentachorda** Meyrick, 1915 (Cosmopteryx [sic]) Peru
22 **pyrozela** Meyrick, 1922 (Cosmopteryx [sic]) Brazil (Pa)
23 **sanctivicentii** Walsingham, 1892 (Cosmopteryx [sic]) St. Vincent
24 **similis** Walsingham, 1897 (Cosmopteryx [sic]) Virgin Is. (St. Croix)
25 **teligera** Meyrick, 1915 (Cosmopteryx [sic]) Colombia
26 **tenax** Meyrick, 1915 (Cosmopteryx [sic]) Colombia
27 **tetragramma** Meyrick, 1915 (Cosmopteryx [sic]) Guyana
28 **thrasyzela** Meyrick, 1915 (Cosmopteryx [sic]) Guyana
29 **venefica** Meyrick, 1915 (Cosmopteryx [sic]) Peru
30 **violenta** (Meyrick, 1916) (Batrachedra) Guyana
31 **xanthura** Walsingham, 1909 (Cosmopteryx [sic]) Mexico (Tab)

DROMIAULIS Meyrick, 1922

32 **excitata** Meyrick, 1922 Peru

ECBALLOGONIA Walsingham, 1912

33 **bimetallica** Walsingham, 1912 Mexico (Ver)

HARPOGRAPTIS Meyrick, 1925

34 **eucharacta** (Meyrick, 1922) (Stomopteryx) Brazil (Pa)

LABDIA Walker, 1864

35 **cyanocoma** Meyrick, 1922 Peru

LIMNAECIA Stainton, 1851
Lymnaecia Kimball, 1965, missp.

36 **dasytricha** Meyrick, 1917 Guyana

MORILOMA Busck, 1912

37 **pardella** Busck, 1912 Panama

PYRODERCES Herrich-Schäffer, 1853
Anatrachyntis Meyrick, 1915
Sathrobrota Hodges, 1962

38 **albistrigella** (Möschler, 1890) (Batrachedra) Puerto Rico
39 **rileyi** (Walsingham, 1882) (Batrachedra) USA (Ga)
stigmatophora (Walsingham, 1897) (Batrachedra) Virgin Is. (St. Thomas)

"**SCAEOSOPHA**" auct. (not Meyrick, 1914)

40 **albicellata** Meyrick, 1931, mispl. Brazil
41 **citrocarpa** Meyrick, 1931, mispl. Brazil (ES)

SEMATOPTIS Meyrick, 1931

42 **amphilychna** Meyrick, 1931 Chile

TELADOMA Busck, 1932

*43 **murina** Hodges, 1962 USA (Az)
*44 **nebula** Hodges, 1978 USA (Az)

TRICLONELLA Busck, 1901
Anorcota Meyrick, 1920
Pharmacoptis Meyrick, 1932

45 **aglaogramma** Meyrick, 1931 Brazil (ES)
46 **antidectis** (Meyrick, 1914) (Epicallima) USA (Az)
*47 **bicoloripennis** Hodges, 1962 USA (Tx)
48 **calyptrodes** Meyrick, 1922 Brazil (Am)
49 **chionozona** Meyrick, 1931 Brazil
50 **cruciformis** Meyrick, 1931 Brazil (ES)
51 **diglypta** Meyrick, 1931 Brazil
52 **elliptica** Meyrick, 1916 French Guiana
53 **etearcha** Meyrick, 1920 Peru
54 **euzosta** Walsingham, 1912 Panama
55 **iphicleia** Meyrick, 1924 Costa Rica
56 **philantha** Meyrick, 1920 Brazil (Am)
57 **pictoria** Meyrick, 1916 French Guiana
58 **platyxantha** (Meyrick, 1909) (Promalactis) Bolivia
59 **rhabdophora** Forbes, 1930 Puerto Rico
breviramis (Meyrick, 1932) (Pharmacoptis) ?
60 **sequella** Busck, 1914 Panama
61 **trachyxyla** Meyrick, 1920 Brazil (Pa)
62 **triargyra** Meyrick, 1920 Peru
63 **turbinalis** Meyrick, 1933 Argentina
64 **umbrigera** Meyrick, 1929 Panama
65 **xanthota** Walsingham, 1912 Mexico (Gue)

URANGELA Busck, 1912

66 **pygmaea** Busck, 1912 Panama

Chrysopeleiinae

ASCALENIA Wocke, 1876

67 **pancrypta** (Meyrick, 1915) (Cholotis) Cuba
68 **plumbata** (Meyrick, 1915) (Cholotis) Guyana
69 **praediata** Meyrick, 1922 Peru
70 **revelata** Meyrick, 1922 Brazil (Pa)
71 **vadata** Meyrick, 1922 Peru

ERITARBES Walsingham, 1909

72 **guttata** Busck, 1914 Panama
73 **otiosa** Walsingham, 1909 Mexico (Tab)

ITHOME Chambers, 1875
Eriphia Chambers, 1875, preocc. (Latreille, 1817 [Crustacea])

74 **anthraceuta** (Meyrick, 1915) (Cholotis) Peru
75 **cathidrota** (Meyrick, 1915) (Cholotis) Guyana
76 **chersota** (Meyrick, 1915) (Cholotis) Colombia
77 **concolorella** (Chambers, 1875) (Eriphia) USA (Tx)
unimaculella Chambers, 1875 USA (Tx)
unomaculella auct., missp.
78 **curvipunctella** (Walsingham, 1892) (Anybia) St. Vincent
quinquepunctata (Forbes, 1931) (Eriphia) Puerto Rico
*79 **edax** Hodges, 1962 USA (Tx)
80 **iresiarcha** (Meyrick, 1915) (Cholotis) Guyana
81 **pelasta** (Meyrick, 1915) (Cholotis) Guyana
82 **pernigrella** (Forbes, 1931) (Eriphia) Puerto Rico
83 **pignerata** (Meyrick, 1922) (Ascalenia) Peru

LEPTOZESTIS Meyrick, 1924

84 **chalcoptila** (Meyrick, 1922) (Syntomactis) Peru

OBITHOME Hodges, 1964

*85 **punctiferella** (Busck, 1906) (Mompha) USA (Tx)

PERIMEDE Chambers, 1874

86 **annulata** Busck, 1914 Panama
*87 **battis** Hodges, 1962 USA (Az)
88 **catapasta** Walsingham, 1909 Mexico (Gue)
89 **purpurascens** Forbes, 1931 Puerto Rico

PERIPLOCA Braun, 1919

*90 **juniperi** Hodges, 1978 USA (Wy)

STILBOSIS Clemens, 1860
Aeaea Chambers, 1874
Amaurogramma Braun, 1919

91 **alcyonis** Meyrick, 1917 French Guiana
92 **alsocoma** Meyrick, 1917 Guyana
93 **amphibola** Walsingham, 1909 Mexico (Gue)
94 **argyritis** Meyrick, 1922 Brazil (Pa)
95 **chrysorrhabda** Meyrick, 1922 Peru
96 **condylota** Meyrick, 1917 Guyana
97 **devoluta** Meyrick, 1917 Guyana
98 **gnomonica** Meyrick, 1917 Guyana
99 **hypanthes** Meyrick, 1917 Guyana
100 **incincta** Walsingham, 1909 Mexico (Tab)
101 **juvenis** Walsingham, 1909 Mexico (Gue)
102 **lonchocarpella** Busck, [1934] Cuba
103 **phaeoptera** Forbes, 1931 Puerto Rico
104 **symphracta** Meyrick, 1917 Guyana
105 **tesquella** Clemens, 1860 USA [Pa?]
tesquatella Chambers, 1878, missp.
quinquicristatella (Chambers, 1881) (Laverna) USA (Ma)

WALSHIA Clemens, 1864

106 **albicornella** Busck, 1914 Panama
107 **calcarata** Walsingham, 1909 Mexico (Tab)
108 **detracta** Walsingham, 1909 Mexico (Gue)
109 **pentapyrga** (Meyrick, 1922) (Mompha) Brazil

28. SCYTHRIDIDAE

by V. O. Becker

SCYTHRIS Hübner, [1825]
Butalis Treitschke, 1833, preocc. (Boie, 1826 [Aves])
Arotrura Walsingham, 1888
Colinita Busck, 1907

1 **depressa** Meyrick, 1931 Paraguay
2 **dimota** Meyrick, 1931 Paraguay
3 **dividua** Meyrick, 1916 Peru
4 **ejiciens** Meyrick, 1928 Peru
5 **fluvialis** Meyrick, 1916 Colombia
6 **immunis** Meyrick, 1916 Peru
7 **inanima** Meyrick, 1916 Peru
8 **medullata** Meyrick, 1916 Peru
9 **notorrhoa** Meyrick, 1921 Brazil (Am)
10 **plocogastra** Meyrick, 1931 Paraguay
11 **tibicina** Meyrick, 1916 Peru
12 **zeugmatica** Meyrick, 1931 Brazil (Pa)

29. GELECHIIDAE [28]

by V. O. Becker

Anomologinae

ISOPHRICTIS Meyrick, 1917

*1 **actiella** Barnes & Busck, 1920 USA (Ca)

MEGACRASPEDUS Zeller, 1839
Neda Chambers, 1874, preocc. (Mulsant, 1850 [Coleoptera])
Pycnobathra Lower, 1901
Autoneda Busck, 1903, repl. name
Toxoceras Chrétien, 1915
Megacraspedas Barnes & McDunnough, 1917, missp.

2 **exilis** Walsingham, 1909 Mexico (Gue)

MONOCHROA Heinemann, 1870
Catabrachmia Rebel, 1909

3 **absconditella** (Walker, 1864) (Gelechia) "N. Am."
palpiannulella (Chambers, 1872) (Gelechia) USA (Ky)

NEALYDA Dietz, 1900

4 **accincta** Meyrick, 1923 Brazil (Pa)
5 **bicolor** (Walsingham, 1892)(Didactylota) St. Vincent
6 **bougainvilleae** E. M. Hering, 1955 Argentina
7 **leucozostra** Meyrick, 1923 Brazil (Pa)
8 **pisoniae** Busck, 1900 USA (Fl)

Gelechiinae

ARISTOTELIA Hübner, [1825]
Ergatis Heinemann, 1870, preocc. (Blackwell, 1841 [Arachnida])
Isochasta Meyrick, 1886
Eucatoptus Walsingham, 1897

9 **aphiltra** Meyrick, 1917 Peru
10 **argyractis** Meyrick, 1923 Brazil (Am)
11 **calculatrix** Meyrick, 1923 Brazil (Pa)
12 **chalybeichroa** (Walsingham, 1897) (Eucatoptus) Virgin Is. (St. Thomas)
chalybochroa Meyrick, 1925, emend.
13 **chloroneura** Meyrick, 1923 Brazil (Pa)
14 **corallina** Walsingham, 1909 Mexico (Gue)
15 **cosmographa** Meyrick, 1917 Peru
16 **crassicornis** Walsingham, 1897 Virgin Is. (St. John)
17 **cynthia** Meyrick, 1917 Peru
18 **cytheraea** Meyrick, 1917 Colombia
19 **dasypoda** Walsingham, 1910 Mexico (Tab)
20 **diolcella** Forbes, 1931 Puerto Rico
21 **elachistella** (Zeller, 1877) (Gelechia) Colombia
22 **ephoria** Meyrick, 1917 Peru
23 **erycina** Meyrick, 1917 Ecuador
24 **eupatoriella** Busck, [1934] Cuba
25 **hieroglyphica** Walsingham, 1909 Mexico (Tab)
26 **howardi** Walsingham, 1909 Mexico (Son)
27 **lignicolora** Forbes, 1931 Puerto Rico
28 **naxia** Meyrick, 1926 Galapagos Is.
29 **oribatis** Meyrick, 1917 Peru
30 **pantalaena** (Walsingham, 1911) (Untomia) Mexico (Tab)
31 **paphia** Meyrick, 1917 Peru
32 **parephoria** Clarke, 1951 Argentina
33 **paterata** Meyrick, 1914 Guyana
34 **penicillata** (Walsingham, 1897) (Eucatoptus) Haiti
35 **perfossa** Meyrick, 1917 Ecuador
radicata Meyrick, 1917 Colombia
36 **perplexa** Clarke, 1951 Argentina
37 **probolopis** Meyrick, 1923 Brazil (Am)
38 **pudibundella** (Zeller, 1873) (Gelechia) USA (Tx)
intermediella (Chambers, 1878) (Gelechia) USA (Tx)
39 **pulicella** Walsingham, 1897 Virgin Is. (St. Thomas)
40 **pyrodercia** Walsingham, 1910 Mexico (Gue)
41 **roseosuffusella** (Clemens, 1860) (Gelechia) USA [Pa?]
belella (Walker, 1864) (Gelechia) "N. Am."
42 **rubidella** (Clemens, 1860) (Gelechia) USA [Pa?]
rubensella (Chambers, 1872) (Gelechia) USA (Ky)
pudibundella.-(Chambers, 1877) (Gelechia), misid. (not Zeller, 1873)
43 **sarcodes** Walsingham, 1910 Panama
44 **saturnina** Meyrick, 1917 Peru
45 **squamigera** Walsingham, 1909 Mexico (Gue)
46 **subrosea** Meyrick, 1914 Guyana
47 **trossulella** Walsingham, 1897 Dom. Rep.
48 **vagabundella** Forbes, 1931 Puerto Rico
49 **veteranella** (Zeller, 1877) (Tachyptilia) Brazil
50 **vicana** Meyrick, 1917 Peru

ARLA Clarke, 1942

*51 **diversella** (Busck, 1916) (Gelechia) USA (Ca)

AROGA Busck, 1914
Aruga Janse, 1958, missp.

52 **paraplutella** (Busck, 1910) (Gelechia) USA (Ca)

AROGALEA Walsingham, 1910

53 **albilingua** Walsingham, 1911 Mexico (Gue)
54 **archaea** Walsingham, 1911 Mexico (Gue)
55 **crocipunctella** (Walsingham, 1892) (Lita) St. Vincent
56 **melitoptila** (Meyrick, 1923) (Telphusa) Brazil (Am)
57 **senecta** Walsingham, 1911 Mexico (Gue)
58 **soronella** Busck, 1914 Panama

BARTICEJA Povolný, 1967

59 **epitricha** (Meyrick, 1917) (Phthorimaea) Guyana
ochrotoma (Meyrick, 1923) (Telphusa) Brazil (Pa)

CALLIPRORA Meyrick, 1914

60 **centrocrossa** Meyrick, 1922 Brazil (Am)
61 **clistogramma** Meyrick, 1926 Brazil (SP)
62 **erethistis** Meyrick, 1922 Peru
63 **eurydelta** Meyrick, 1922 Peru
64 **pentagramma** Meyrick, 1914 Guyana
65 **peritura** Meyrick, 1922 Brazil (Pa)
66 **platyxipha** Meyrick, 1922 Brazil (Am)
67 **rhodogramma** Meyrick, 1922 Brazil (Am)
68 **tetraplecta** Meyrick, 1922 Peru
69 **trigramma** Meyrick, 1914 Guyana

CHIONODES Hübner, [1825]
Chionoda Hübner, [1826], missp.

70 **argosema** (Meyrick, 1917) (Gelechia) Ecuador
71 **consona** (Meyrick, 1917) (Gelechia) Peru
72 **dryobathra** (Meyrick, 1917) (Gelechia) Colombia
73 **eburata** (Meyrick, 1917) (Gelechia) Colombia
74 **icriodes** (Meyrick, 1931) (Gelechia) Chile
75 **lacticoma** (Meyrick, 1917) (Gelechia) Peru
76 **litigiosa** (Meyrick, 1917) (Gelechia) Ecuador
77 **mediofuscella** (Clemens, 1863) (Gelechia) USA [Pa?]
vagella (Walker, 1864) (Gelechia) Canada
fuscoochrella (Chambers, 1872) (Depressaria) USA (Ky)
liturosella (Zeller, 1873) (Gelechia) USA (Tx)
rhedaria (Meyrick, 1923) (Gelechia) Canada
78 **pentadora** (Meyrick, 1917) (Gelechia) French Guiana
79 **perissosema** (Meyrick, 1932) (Gelechia) Argentina
80 **salva** (Meyrick, 1925) (Phthorimaea), repl. name Virgin Is. (St. Thomas)
leucocephala (Walsingham, 1897) (Gelechia), preocc. (not Lower, 1893) Virgin Is. (St. Thomas)
81 **spiridoxa** (Meyrick, 1931) (Gelechia) Paraguay

COLEOTECHNITES Chambers, 1880
Evagora Clemens, 1860, preocc. (Péron & Lesueur, 1810 [Coelenterata])
Eidothea Chambers, 1873, preocc. (Risso, 1826 [Mollusca])
Eidothoa Chambers, 1873, missp.
Coleotechnistes Riley, 1891, missp.
Eucordylea Dietz, 1900
Pulicalvaria T. N. Freeman, 1963

*82 **elucidella** (Barnes & Busck, 1920) (Eucordylea) USA (Ca)
83 **vagatioella** (Chambers, 1873) (Eidothoa [sic]) USA (Ky)
dorsivittella (Zeller, 1873) (Gelechia) USA (Tx)

DELTOPHORA Janse, 1950

84 **caymana** Sattler, 1979 Cayman Is.
85 **duplicata** Sattler, 1979 Cayman Is.
86 **flavocincta** Sattler, 1979 Colombia
87 **glandiferella** (Zeller, 1873) (Gelechia) USA (Tx)
88 **lanceella** Sattler, 1979 Guyana
89 **minuta** Sattler, 1979 Brazil (Pa)
90 **suffusella** Sattler, 1979 Paraguay

EPHYSTERIS Meyrick, 1908
Microcraspedus Janse, 1958
Ephystereris Janse, 1960, missp.
Echinoglossa Clarke, 1965

91 **trinota** (Clarke, 1965) (Echinoglossa) Chile (Juan Fernandez Is.)

EUCHIONODES Clarke, 1950

92 **traditionis** Clarke, 1950 Argentina

EUDACTYLOTA Walsingham, 1911

93 **barberella** (Busck, 1903) (Neodactylota) USA (Az)
94 **iobapta** (Meyrick, 1927) (Stomopteryx) USA (Tx)

EVIPPE Chambers, 1873
Phaetusa Chambers, 1875, preocc. (Wagler, 1832 [Aves])
Tholerostola Meyrick, 1917

95 **aequorea** (Meyrick, 1917) (Recurvaria) Peru
96 **aulonota** (Meyrick, 1917) (Aristotelia) Ecuador
97 **evippella** (Forbes, 1931) (Tholerostola) Puerto Rico
98 **leuconota** (Zeller, 1873) (Gelechia) USA (Tx)
plutella (Chambers, 1875) (Phaetusa) USA (Tx)
99 **omphalopa** (Meyrick, 1917) (Tholerostola) Ecuador
100 **plumata** (Meyrick, 1917) (Aristotelia) Guyana

EXOTELEIA Wallengren, 1881
Paralechia Busck, 1903
Heringia Spuler, 1910, preocc. (Rondani, 1856 [Diptera])
Heringiola Strand, 1917, repl. name

101 **ithycosma** (Meyrick, 1914) (Strobisia) Guyana

FACULTA Busck, 1939

102 **inaequalis** (Busck, 1910) (Gelechia) — USA (NM)
inaequalis (Walsingham, 1911) (Gelechia), preocc. (not Busck, 1910) — Mexico (Son)
anisectis (Meyrick, 1923) (Gelechia), repl. name — Mexico (Son)
clistrodoma (Meyrick, 1923) (Gelechia) — USA (Az)

FASCISTA Busck, 1939

103 **albipectus** (Walsingham, 1911) (Gelechia) — Mexico (Son)
104 **vorax** (Meyrick, 1939) (Phthorimaea) — Argentina

FRISERIA Busck, 1939

105 **acaciella** (Busck, 1906) (Telphusa) — USA (Tx)
*106 **caieta** Hodges, 1966 — USA (Az)
107 **cockerelli** (Busck, 1903) (Gelechia) — USA (NM)
lindenella (Busck, 1903) (Gelechia) — USA (Tx)
malindella (Busck, 1910) (Gelechia) — USA (NM)
sarcochlora (Meyrick, 1929) (Gelechia) — USA (NM)
108 **infracta** (Walsingham, 1911) (Gelechia) — Mexico (Gue)
109 **lacticaput** (Walsingham, 1911) (Gelechia) — Mexico (Gue)
lacticeps (Meyrick, 1925), emend.
*110 **nona** Hodges, 1966 — USA (Az)
111 **paphlactis** (Meyrick, 1912) (Gelechia) — Brazil (SP)
112 **repentina** (Walsingham, 1911) (Gelechia) — Mexico (Gue)

GELECHIA Hübner, [1825]
Guenea Bruand, 1850
Galechia Desmarest, [1857], missp.
Gelschia Nowicki, 1865, missp.
Cirrha Chambers, 1872
Oeseis Chambers, 1875
Gelecia Watt, 1920, missp.

113 **bathrochlora** Meyrick, 1932 — Brazil (SC)
114 **bufo** Walsingham, 1911 — Mexico (Gue)
115 **cacoderma** Walsingham, 1911 — Mexico (Gue)
116 **caespitella** Zeller, 1877 — Colombia
117 **cerussata** Walsingham, 1911 — Mexico (Ver)
118 **chlorocephala** Meyrick, 1932 — Mexico (DF)
119 **clopica** Meyrick, 1931 — Argentina
120 **concinna** Walsingham, 1911 — Mexico (Gue)
121 **creberrima** Walsingham, 1911 — Mexico (Gue)
122 **cuneifera** Walsingham, 1911 — Mexico (Gue)
123 **delapsa** Meyrick, 1931 — Brazil (ES)
124 **diacmota** Meyrick, 1932 — Brazil (SC)
125 **dolbyi** (Walsingham, 1911) (Dichomeris) — Panama
126 **elephantopis** Meyrick, 1936 — Venezuela
127 **exclarella** Möschler, 1890 — Puerto Rico
128 **flammulella** Walsingham, 1897 — Virgin Is. (St. Thomas)
129 **gnathodoxa** Meyrick, 1926 — Galapagos Is.
130 **goniospila** Meyrick, 1931 — Argentina
131 **hetaeria** Walsingham, 1911 — Mexico (Ver)
132 **impurgata** Walsingham, 1911 — Mexico (Son)
133 **lapidescens** Meyrick, 1916, repl. name — Mexico (Son)
lithodes Walsingham, 1911, preocc. (not Meyrick, 1886) — Mexico (Son)
134 **leptospora** Meyrick, 1932 — Costa Rica
135 **nephelophracta** Meyrick, 1932 — Costa Rica
136 **neptica** Walsingham, 1911 — Mexico (Gue)
137 **nigripectus** Walsingham, 1911 — Mexico (Son)
138 **nucifer** Walsingham, 1911 — Mexico (Son)
nucifera Meyrick, 1925, emend.
139 **ophiaula** Meyrick, 1931 — Argentina
140 **ophiomorpha** Meyrick, 1935 — Argentina
141 **pertinens** Meyrick, 1931 — Paraguay
142 **petraea** Walsingham, 1911 — Guatemala
143 **picrogramma** Meyrick, 1929 — Guyana
144 **platydoxa** Meyrick, 1923 — Guyana
145 **pleroma** Walsingham, 1911 — Mexico (Gue)
146 **protozona** Meyrick, 1926 — Galapagos Is.
147 **rhypodes** Walsingham, 1911 — Mexico (Son)
148 **scotodes** Walsingham, 1911 — Mexico (Son)
149 **sonorensis** Walsingham, 1911 — Mexico (Son)
150 **suspensa** Meyrick, 1923 — Brazil (Am)
151 **synthetica** Walsingham, 1911 — Mexico (Son)
*152 **thymiata** (Meyrick, 1929) (Nothris) — USA (Az)
153 **traducella** Busck, 1914 — Panama
154 **veneranda** Walsingham, 1911 — Mexico (Son)
155 **xylobathra** Meyrick, 1936 — Venezuela

GNORIMOSCHEMA Busck, 1900
Gnorimochema Dyar, [1903], missp.
Tuta Kieffer & Jörgensen, 1910
Gonorimoschema Deurs, 1954, missp.
Gnorrimoschema Hartig, 1964, missp.
Lerupsia Riedl, 1965
Neoschema Povolný, 1967
Larupsia Soffner, 1967, missp.

156 **atriplicella** Kieffer & Jörgensen, 1910 — Argentina
157 **borsaniella** Köhler, 1939 — Argentina
158 **cestrivora** Clarke, 1950 — Argentina
cestivora Hayward, 1969, missp.
159 **dudiella** Busck, 1903 — USA (Az)
160 **euchthonia** (Meyrick, 1939) (Phthorimaea) — Argentina
161 **exacta** (Meyrick, 1917) (Phthorimaea) — Guyana
162 **involuta** (Meyrick, 1917) (Phthorimaea) — Guyana
163 **motasi** Povolný, 1976 — Colombia
164 **perfidiosa** (Meyrick, 1917) (Phthorimaea) — Colombia
165 **saphirinella** (Chambers, 1875) (Gelechia) — USA (Co)
166 **urosema** (Meyrick, 1917) (Phthorimaea) — Peru

HAPALOSARIS Meyrick, 1917

167 **petulans** Meyrick, 1917 — Colombia

KEIFERIA Busck, 1939

168 **brunnea** Povolný, 1973 — Virgin Is. (St. Thomas)
169 **colombiana** Povolný, 1975 — Colombia
170 **lycopersicella** (Walsingham, 1897) (Eucatoptus) — Virgin Is. (St. Croix)
lenta (Meyrick, 1917) (Phthorimaea) — Peru
lycopersicella (Busck, 1928) (Phthorimaea) — USA (Hi)
elmorei (Keifer, 1936) (Gnorimoschema) — USA (Ca)
171 **rusposoria** Povolný, 1970 — Grenada

LITA Treitschke, 1833

172 **dialis** Hodges, 1966 — USA (Az)
*173 **pagella** Hodges, 1966 — USA (Az)
*174 **princeps** (Busck, 1910) (Gnorimoschema) — USA (Ca)
*175 **recens** Hodges, 1966 — USA (Ca)
*176 **rectistrigella** (Barnes & Busck, 1920) (Gelechia) — USA (Ca)
176.1 **texanella** (Chambers, 1880) (Anesychia) — USA (Tx)
chambersella Dyar, [1903] — USA (Tx)
177 **variabilis** (Busck, 1903) (Gelechia) — USA (Ca/Co)

LOCHARCA Meyrick, 1923

178 **emicans** Meyrick, 1923 — Peru

NEODACTYLOTA Busck, 1903

*179 **basilica** Hodges, 1966 — USA (Az)

ORSOTRICHA Meyrick, 1914

180 **venosa** (Butler, 1883) (Topeutis) — Chile

PHTHORIMAEA Meyrick, 1902
Phtyorimaea Turner, 1919, missp.
Phthorimoea Povolný & Zakopal, 1951, missp.
Pthorimaea Issiki, 1957, missp.
Phthorimea Diakonoff, [1968], missp.
Phtorimea Oei-Dharma, 1969, missp.

181 **ferella** (Berg, 1875) (Gelechia) — Argentina
182 **impudica** Walsingham, 1911 — Panama
183 **interjuncta** Meyrick, 1931 — Brazil
184 **jamaicensis** (Walsingham, 1897) (Gelechia) — Jamaica
185 **operculella** (Zeller, 1873) (Gelechia) — USA (Tx)
terrella (Walker, 1864) (Gelechia), preocc. (not Denis & Schiffermüller, 1775) — Australia
solanella (Boisduval, 1875) (Bryotropha) — ?
tabacella (Ragonot, 1879) (Gelechia) — ?
sedata (Butler, 1880) (Gelechia) — New Zealand
epicentra Meyrick, 1909 — South Africa
186 **sphenophora** (Walsingham, 1897) (Gelechia) — Grenada

POLYHYMNO Chambers, 1874
Copocercia Zeller, 1877
Oegoconiodes Matsumura, 1931
Oegoconides Neave, 1940, missp.

187 **charigramma** Meyrick, 1929 — Brazil (Am)
188 **colleta** Walsingham, 1911 — Mexico (Gue)
189 **conflicta** Meyrick, 1917 — Peru
190 **convergens** Walsingham, 1911 — Mexico (Gue)
191 **crambinella** (Zeller, 1877) (Copocercia) — Colombia
192 **gladiata** Meyrick, 1917 — Colombia
193 **leucocras** Walsingham, 1911 — Mexico (Son)
194 **luteostrigella** Chambers, 1874 — USA (Tx)
fuscostrigella Chambers, 1876 — USA (Tx)
195 **subaequalis** Walsingham, 1911 — Mexico (Gue)

PSEUDARLA Clarke, 1965

196 **miranda** Clarke, 1965 — Chile (Juan Fernandez Is.)

PTYCERATA Ely, 1910
Scrobipalpopsis Povolný, 1967
Subgenus **Ptycerata** Ely, 1910

197 **chili** (Povolný, 1967) (Scrobipalpopsis) Chile
198 **solanivora** (Povolný, 1973) (Scrobipalpopsis) Costa Rica

Subgenus **Scrobischema** Povolný, 1979

199 **vergarai** (Povolný, 1979) (Scrobipalpopsis) Colombia

RECURVARIA Haworth, 1828
Telea Stephens, 1834, preocc. (Hübner, 1819 [Saturniidae])
Aphanaula Meyrick, 1895
Hinnebergia Spuler, 1910
Microlechia Turati, 1924

200 **annulicornis** (Walsingham, 1897) (Aristotelia) Virgin Is. (St. Thomas)
201 **eromene** (Walsingham, 1897) (Aristotelia) Virgin Is. (St. Thomas)
202 **febriculella** (Zeller, 1877) (Teleia) Colombia
203 **filicornis** (Zeller, 1877) (Teleia) Colombia
204 **flagellifer** Walsingham, 1910 Mexico (Gue)
flagellifera Meyrick, 1925, missp.
205 **insequens** Meyrick, 1931 Brazil (SP)
206 **intermissella** (Zeller, 1877) (Teleia) Colombia
207 **kitella** (Walsingham, 1897) (Aristotelia) Haiti
208 **melanostictella** (Zeller, 1877) (Teleia) Colombia
209 **merismatella** (Zeller, 1877) (Teleia) Colombia
210 **nothostigma** Meyrick, 1914 Guyana
211 **ornatipalpella** (Walsingham, 1897) (Aristotelia) Grenada
212 **ostariella** (Walsingham, 1897) (Aristotelia) Virgin Is. (St. Thomas)
213 **penetrans** Meyrick, 1923 Brazil (Am)
214 **picula** Walsingham, 1910 Mexico (Ver)
215 **pleurosaris** Meyrick, 1923 Brazil (Am)
216 **putella** Busck, 1914 Panama
217 **rhicnota** Walsingham, 1910 Guatemala
218 **rhombophorella** (Zeller, 1877) (Teleia) Colombia
219 **sartor** Walsingham, 1910 Mexico (Gue)
220 **saxea** Meyrick, 1923 Brazil (Pa)
221 **senariella** (Zeller, 1877) (Teleia) Colombia
222 **sticta** Walsingham, 1910 Mexico (Gue)
223 **synestia** Meyrick, 1939 Argentina
224 **thiodes** Meyrick, 1917 Colombia
225 **thysanota** Walsingham, 1910 Mexico (Gue)
226 **trigonophorella** (Zeller, 1877) (Teleia) Colombia
227 **xanthotricha** Meyrick, 1917 Peru

SCHISTOPHILA Chrétien, 1899

228 **fuscella** Forbes, 1931 Puerto Rico

SCROBIPALPULA Povolný, 1964
Subgenus **Eurysacca** Povolný, 1967

229 **melanocampa** (Meyrick, 1917) (Phthorimaea) Peru

Subgenus **Magnifascia** Povolný, 1967

230 **aulorrhoa** (Meyrick, 1935) (Phthorimaea) Argentina

Subgenus **Scrobipalpula** Povolný, 1964

231 **absoluta** (Meyrick, 1917) (Phthorimaea) Peru
232 **chiquitella** (Busck, 1909) (Gnorimoschema) USA (NM)
233 **conifera** (Meyrick, 1916) (Chelaria) Ecuador
234 **crustaria** (Meyrick, 1917) (Phthorimaea) Peru
235 **daturae** (Zeller, 1877) (Doryphora) Colombia
236 **densata** (Meyrick, 1917) (Phthorimaea) Peru
laciniosa (Meyrick, 1931) (Phthorimaea) Argentina
237 **gregalis** (Meyrick, 1917) (Phthorimaea) Peru
238 **gregariella** (Zeller, 1877) (Lita) Colombia
239 **hemilitha** (Clarke, 1965) (Gnorimoschema) Chile (Juan Fernandez Is.)
240 **henshawiella** (Busck, 1903) (Gnorimoschema) USA (Co)
ochreostrigella (Chambers, 1877) (Gelechia), preocc. (not Chambers, 1875) USA (Co)
241 **ilyella** (Zeller, 1877) (Lita) Colombia
242 **isochlora** (Meyrick, 1931) (Phthorimaea) Paraguay
243 **melanolepis** (Clarke, 1965) (Gnorimoschema) Chile (Juan Fernandez Is.)
244 **motasi** Povolný, 1976 Colombia
245 **patagonica** Povolný, 1977 Argentina
246 **stirodes** (Meyrick, 1931) (Phthorimaea) Argentina
247 **trichinaspis** (Meyrick, 1917) (Phthorimaea) Peru

SOPHRONIA Hübner, [1825]

248 **mediatrix** Zeller, 1877 Colombia

STEGASTA Meyrick, 1904

249 **biniveipunctata** (Walsingham, 1897) (Gelechia) Grenada
250 **bosqueella** (Chambers, 1875) (Oecophora) USA
basqueella (Chambers, 1875) (Oecophora), missp.
bosquella (Chambers, 1878) (Gelechia), emend.
costipunctella (Möschler, 1890) (Gelechia) Puerto Rico
251 **capitella** (Fabricius, 1794) (Alucita) "West Indies"
capitatus (Fabricius, 1798) (Ypsolophus), repl. name "West Indies"
robustella (Walker, 1864) (Gelechia) Dom. Rep.
rivulella (Möschler, 1890) (Gelechia) Puerto Rico
252 **comissata** Meyrick, 1923 Brazil (Am)
253 **donatella** (Walker, 1864) (Gelechia) Jamaica
254 **phalacra** (Walsingham, 1911) (Gelechia) Mexico (Son)
255 **postpallescens** (Walsingham, 1897) (Gelechia) Grenada
256 **scoteropis** Meyrick, 1931 Paraguay
257 **zygotoma** Meyrick, 1917 Ecuador

STOMOPTERYX Heinemann, 1870
Inotica Meyrick, 1913
Instica Sharp, 1915, missp.
Acraeologa Meyrick, 1921
Stomopterix Turati, 1922, missp.
Stromopteryx Pierce & Metcalfe, 1935, missp.

258 **phaeopa** Meyrick, 1918 Peru

SYMMETRISCHEMA Povolný, 1967

259 **altisona** (Meyrick, 1917) (Phthorimaea) Peru
260 **aquilina** (Meyrick, 1917) (Phthorimaea) Peru
261 **ardeola** (Meyrick, 1931) (Phthorimaea) Paraguay
262 **atrifascis** (Meyrick, 1917) (Phthorimaea) Peru
263 **capsica** (Bradley & Povolný, 1965) (Gnorimoschema) Montserrat
264 **capsivorum** Povolný, 1973 Peru
265 **loquax** (Meyrick, 1917) (Phthorimaea) Peru
266 **plaesiosema** (Turner, 1919) (Phthorimaea) Australia
melanoplintha (Meyrick, 1926) (Phthorimaea) New Zealand
tuberosella (Busck, 1931) (Gnorimoschema) Peru
267 **striatella** (Murtfeldt, 1900) (Eucatoptus) USA (Mo)
268 **ventratella** (Zeller, 1877) (Gelechia) Colombia

TAYGETE Chambers, 1873

269 **altivola** (Meyrick, 1929) (Epithectis) Peru
270 **balsamopa** (Meyrick, 1923) (Epithectis) Peru
271 **barydelta** (Meyrick, 1923) (Epithectis) Brazil (Pa)
272 **citranthes** (Meyrick, 1923) (Epithectis) Brazil (Pa)
273 **consociata** (Meyrick, 1914) (Epithectis) Guyana
274 **critica** (Walsingham, 1910) (Epithectis) Mexico (Gue)
275 **ignavella** (Zeller, 1877) (Gelechia) Colombia
276 **lasciva** (Walsingham, 1910) (Epithectis) Panama
277 **notospila** (Meyrick, 1923) (Epithectis) Brazil (Am)
278 **parvella** (Fabricius, 1794) (Alucita) "West Indies"
xylochroa (Meyrick, 1926) (Epithectis) Bermuda
279 **platysoma** (Walsingham, 1910) (Epithectis) Mexico (Ver/Dur)
280 **sphecophila** (Meyrick, 1936) (Epithectis) Trinidad

TELPHUSA Chambers, 1872
Adrasteia Chambers, 1872
Adrastia Kirby, 1874, emend.
Geniadophora Walsingham, 1897
Telephusa Beirne, 1938, missp.

281 **auxoptila** Meyrick, 1926 Colombia
282 **callitechna** Meyrick, 1914 Guyana
praefinita (Meyrick, 1917) (Mompha) French Guiana
283 **delatrix** Meyrick, 1923 Peru
284 **distictella** Forbes, 1931 Puerto Rico
285 **extranea** (Walsingham, 1892) (Poecilia) St. Vincent
286 **hemicycla** Meyrick, 1932 Paraguay
287 **latebricola** Meyrick, 1932 Virgin Is. (St. Thomas)
288 **medulella** Busck, 1914 Panama
289 **melanoleuca** Walsingham, 1911 Mexico (Gue)
290 **obligata** Busck, 1914 Panama
291 **ochrifoliata** Walsingham, 1911 Mexico (Ver)
292 **orgilopis** Meyrick, 1923 Brazil (Pa)
293 **penetratrix** Meyrick, 1931 Paraguay
294 **perspicua** (Walsingham, 1897) (Gelechia) Haiti
295 **quinquedentata** (Walsingham, 1911) (Gelechia) Mexico (Gue)
296 **ripula** Walsingham, 1911 Guatemala
297 **smaragdopis** Meyrick, 1926 Costa Rica
298 **translucida** (Walsingham, 1892) (Bryotropha) Dominica

THIOTRICHA Meyrick, 1886
Reuttia O. Hofmann, 1898
Thiotrica Inoue, 1954, missp.
Thiothricha Hartig, 1956, missp.

299 **argoxantha** Meyrick, 1914 Guyana
300 **aucupatrix** Meyrick, 1929 Peru
301 **cleodorella** (Zeller, 1877) (Gelechia) Colombia
302 **godmani** (Walsingham, 1892) (Polyhymno) St. Vincent
303 **laterestriata** (Walsingham, 1897) (Polyhymno) Virgin Is. (St. Thomas)
304 **sciurella** (Walsingham, 1897) (Polyhymno) Virgin Is. (St. Thomas)

TILDENIA Povolný, 1967

305 **gudmannella** (Walsingham, 1897) (Gelechia) Virgin Is. (St. Thomas)

Anacampsinae

ANACAMPSIS Curtis, 1827
Anacompsis Desmarest, [1857], missp.
Tachyptilia Heinemann, 1870
Tuchyptilia Kirby, 1871, missp.
Tachiptilia Chambers, 1878, missp.
Tachyptilix Hartmann, 1880, missp.
Agriastis Meyrick, 1914
Agriastsi Busck, 1919, missp.
Tachoptilia Daltry, 1926, missp.
Trachyphilia Le Marchand, 1947, missp.
Trachyptilia Le Marchand, 1947, missp.

306	**aedificata** Meyrick, 1929	Brazil (Pa)
307	**capyrodes** Meyrick, 1922	Brazil (Am)
308	**cenelpis** (Walsingham, 1911) (Untomia)	Mexico (Tab)
309	**cistulata** Meyrick, 1914	Guyana
310	**conistica** Walsingham, 1910	Mexico (Son)
311	**considerata** Meyrick, 1922	Brazil (Am)
312	**cornifer** Walsingham, 1897	Virgin Is. (St. Croix)
	cornifera Meyrick, 1925, missp.	
313	**crypticopa** (Meyrick, 1931) (Gelechia)	Argentina
314	**diplodelta** Meyrick, 1922	Brazil (Pa)
315	**flexiloqua** Meyrick, 1922	Peru
316	**humilis** Hodges, 1970	Uruguay
317	**idiocentra** Meyrick, 1922	Brazil (Pa)
318	**inquieta** (Meyrick, 1914) (Agriastis)	Guyana
319	**insularis** Walsingham, 1897	Virgin Is. (St. Thomas)
320	**languens** Meyrick, 1918	Ecuador
321	**lapidella** Walsingham, 1897	Grenada
322	**lithodelta** Meyrick, 1922	Peru
323	**meibomiella** Forbes, 1931	Puerto Rico
324	**multinotata** (Meyrick, 1918) (Gelechia)	Guyana
325	**nocturna** (Meyrick, 1914) (Agriastis)	Guyana
326	**peloptila** (Meyrick, 1914) (Agriastis)	Guyana
327	**perquisita** Meyrick, 1922	Brazil (Pa)
328	**petrographa** Meyrick, 1922	Brazil (Pa)
329	**phytomiella** Busck, 1914	Panama
330	**poliombra** Meyrick, 1922	Brazil (Am)
331	**pomaceella** (Walker, 1864) (Gelechia)	Brazil (Am)
332	**prasina** Meyrick, 1914	Guyana
333	**primigenia** Meyrick, 1918	Ecuador
324	**quinquepunctella** Walsingham, 1897	Grenada
335	**rhabdodes** Walsingham, 1910	Mexico (Tab)
336	**scalata** (Meyrick, 1914) (Agriastis)	Guyana
	caneodes Meyrick, 1922	Brazil (Pa)
337	**subactella** (Walker, 1864) (Gelechia)	"Australia" [S. Am.?]
338	**tridentella** (Walsingham, 1910) (Cathegesis)	Mexico (Gue)
339	**ursula** Walsingham, 1910	Mexico (Mor)
340	**viretella** (Zeller, 1877) (Gelechia)	Brazil

BATTARISTIS Meyrick, 1914
Duvita Busck, 1916

341	**acroglypta** Meyrick, 1929	Colombia
342	**amphiscolia** Meyrick, 1914	Guyana
343	**ardiophora** Meyrick, 1914	Guyana
344	**atelesta** Meyrick, 1914	Guyana
345	**bistrigella** (Busck, 1914) (Anacampsis)	Panama
346	**concisa** Meyrick, 1929	Cuba
347	**coniosema** Meyrick, 1922	Peru
348	**curtella** (Busck, 1914) (Anacampsis)	Panama
349	**emissurella** (Walker, 1864) (Gelechia)	Brazil (Am)
	severella (Walker, 1864) (Cryptolechia)	Brazil (Am)
	fuliginosa (R. Felder & Rogenhofer, 1875) (Gelechia)	Brazil (Am)
	dorsalis (Busck, 1914) (Anacampsis)	Panama
	astroconis (Meyrick, 1918) (Compsolechia)	Surinam
350	**ichnota** Meyrick, 1914	Guyana
351	**melanamba** Meyrick, 1914	Guyana
352	**nigratomella** (Clemens, 1863) (Gelechia)	USA [Pa?]
	apicilinella (Clemens, 1863) (Gelechia)	USA [Pa?]
	apicistrigella (Chambers, 1872) (Parasia)	USA (Ky)
353	**orthocampta** Meyrick, 1914	Guyana
354	**parazela** Meyrick, 1929	Brazil (Pa)
355	**perinaeta** (Walsingham, 1910) (Anacampsis)	Mexico (Gue)
356	**prismatopa** Meyrick, 1914	Guyana
357	**rhythmodes** Meyrick, 1929	Brazil (Pa)
358	**sphenodelta** Meyrick, 1922	Brazil (Am)
359	**stereogramma** Meyrick, 1914	Guyana
360	**symphora** (Walsingham, 1911) (Untomia)	Mexico (Tab)
361	**syngraphopa** Meyrick, 1922	Peru
	syngraphora Meyrick, 1925, missp.	
362	**synocha** Meyrick, 1922	Peru
363	**tricentrota** Meyrick, 1931	Brazil
364	**unistrigella** (Busck, 1914) (Anacampsis)	Panama

COMPSOLECHIA Meyrick, 1918

365	**abolitella** (Walker, 1864) (Gelechia)	"Australia" [S. Am.]
366	**abruptella** (Walker, 1864) (Gelechia)	Brazil (Am)
	sectella (Walker, 1864) (Gelechia)	Brazil (Am)
367	**accinctella** (Walker, 1864) (Gelechia)	Brazil (Am)
368	**acosmeta** (Walsingham, 1910) (Anacampsis)	Mexico (Son)
369	**aequilibris** Meyrick, 1931	Brazil (RJ)
370	**amaurota** (Meyrick, 1914) (Anacampsis)	Guyana
371	**amazonica** Meyrick, 1918, repl. name	Brazil (Am)
	suffusella (Walker, 1864) (Gelechia), preocc. (not Douglas, 1850)	Brazil (Am)
372	**ambusta** (Walsingham, 1910) (Anacampsis)	Mexico (Tab)
	brochospila (Meyrick, 1914) (Anacampsis)	Guyana
373	**anthracura** (Meyrick, 1914) (Anacampsis)	Guyana
374	**antiplaca** Meyrick, 1922	Peru
375	**argyracma** Meyrick, 1922	Brazil (Pa)
376	**atmastra** Meyrick, 1929	Peru
377	**balia** (Walsingham, 1910) (Anacampsis)	Mexico (Gue)
378	**binotatella** (Walker, 1864) (Gelechia)	Brazil (Am)
379	**blepharopa** (Meyrick, 1914) (Anacampsis)	Guyana
380	**compalea** (Walsingham, 1910) (Anacampsis)	Mexico (Gue)
381	**canofusella** (Walker, 1864) (Gelechia)	Brazil (Am)
382	**caryoterma** Meyrick, 1922	Brazil (Am)
383	**cassidata** (Meyrick, 1914) (Anacampsis)	Guyana
384	**chelidonia** Meyrick, 1922	Brazil (Pa)
385	**chrysoplaca** (Meyrick, 1912) (Anacampsis)	Venezuela
386	**cognatella** (Walker, 1864) (Gelechia)	Brazil (Am)
387	**corymbas** Meyrick, 1918	Guyana
388	**crocodilopa** Meyrick, 1922	Peru
389	**desectella** (Zeller, 1877) (Tachyptilia)	Cuba
390	**dicax** (Meyrick, 1914) (Anacampsis)	Guyana
391	**diortha** (Meyrick, 1914) (Anacampsis)	Guyana
392	**diplolychna** Meyrick, 1922	Brazil (Pa)
393	**drachmaea** Meyrick, 1922	Brazil (Am)
394	**dryochrossa** Meyrick, 1922	Brazil (Am)
395	**elephas** (Walsingham, 1910) (Anacampsis)	Mexico (Gue)
396	**epibola** (Walsingham, 1910) (Anacampsis)	Mexico (Ver)
397	**erebodelta** Meyrick, 1922	Peru
398	**eupecta** (Meyrick, 1914) (Anacampsis)	Guyana
399	**eurygypsa** Meyrick, 1922	Peru
400	**fasciella** (R. Felder & Rogenhofer, 1875) (Gelechia)	Brazil
401	**ferreata** (Meyrick, 1914) (Anacampsis)	Guyana
402	**glaphyra** (Walsingham, 1910) (Anacampsis)	Mexico (Ver)
403	**halmyra** (Meyrick, 1914) (Anacampsis)	Guyana
404	**hemileucas** Meyrick, 1922	Brazil (Am)
405	**incurva** (Meyrick, 1914) (Anacampsis)	Guyana
406	**inusta** (Meyrick, 1914) (Anacampsis)	Guyana
407	**ischnoptera** Meyrick, 1922	Brazil (Pa)
408	**laguncularieila** (Busck, 1900) (Anacampsis)	USA (Fl)
409	**leucorrhapta** (Meyrick, 1914) (Anacampsis)	Guyana
410	**lingulata** Meyrick, 1918	Colombia
411	**lithomorpha** (Meyrick, 1914) (Anacampsis)	Guyana
412	**loxogramma** Meyrick, 1922	Brazil (Am)
413	**mangelivora** (Walsingham, 1897) (Anacampsis)	Virgin Is. (St. Thomas)
414	**melanophaea** (Forbes, 1931) (Anacampsis)	Puerto Rico
415	**mesodelta** Meyrick, 1922	Brazil (Am)
416	**metadupa** (Walsingham, 1910) (Anacampsis)	Mexico (Tab)
417	**mniocosma** Meyrick, 1922	Peru
418	**molybdina** (Walsingham, 1910) (Anacampsis)	Mexico (Gue)
419	**monochromella** (Walker, 1864) (Gelechia)	Brazil (Am)
	displicitella (Walker, 1864) (Gelechia)	Brazil (Am)
420	**neurophora** Meyrick, 1922	Brazil (Am)
421	**niobella** (R. Felder & Rogenhofer, 1875) (Gelechia)	Brazil (Am)
	tornoptila Meyrick, 1922	Brazil (Am)
422	**niphocentra** Meyrick, 1922	Peru
423	**nuptella** (R. Felder & Rogenhofer, 1875) (Gelechia)	Brazil (Am)
	scholias Meyrick, 1922	Peru
424	**ocelligera** (Butler, 1883) (Gelechia)	Chile
425	**orthophracta** (Meyrick, 1914) (Anacampsis)	Guyana
426	**parmata** Meyrick, 1918	Colombia
427	**peculella** (Busck, 1914) (Anacampsis)	Panama
428	**pentastra** Meyrick, 1922	Brazil (Pa)
429	**percnospila** (Meyrick, 1914) (Anacampsis)	Guyana
430	**perlatella** (Walker, 1864) (Gelechia)	Brazil (Am)
	secundella (Walker, 1864) (Gelechia)	Brazil (Am)
	smaragdulella (Walker, 1864) (Gelechia)	Brazil (Pa)
431	**petromorpha** Meyrick, 1922	Peru
432	**phaeotoxa** Meyrick, 1922	Brazil (Pa)
433	**phepsalitis** Meyrick, 1922	Brazil (Am)
434	**picticornis** (Walsingham, 1897) (Aristotelia)	Virgin Is. (St. Croix)
435	**platiastis** Meyrick, 1922	Brazil (Pa)
436	**plumbeolata** (Walsingham, 1897) (Anacampsis)	Virgin Is. (St. Croix)
437	**praenivea** (Meyrick, 1914) (Anacampsis)	Guyana
438	**ptochogramma** Meyrick, 1922	Brazil (Pa)
439	**pungens** Meyrick, 1922	Peru
440	**quadrifascia** (Walker, 1864) (Gelechia)	Brazil (Am)
	superella (Walker, 1864) (Gelechia)	Brazil (Am)
441	**recta** Meyrick, 1922	Brazil (Am)
442	**refracta** (Meyrick, 1914) (Anacampsis)	Guyana
443	**religata** Meyrick, 1922	Peru
444	**repandella** (Walker, 1864) (Gelechia)	Brazil (Am)
	subscriptella (Walker, 1864) (Gelechia)	Brazil (Am)
	episema (Walsingham, 1910) (Anacampsis)	Mexico (Ver)
445	**rhombica** Meyrick, 1922	Peru
446	**salebrosa** Meyrick, 1918	Colombia
447	**sciomima** Meyrick, 1922	Brazil (Am)
448	**scitella** (Walker, 1864) (Gelechia)	Brazil (Am)

449 **scopulata** (Meyrick, 1914) (Anacampsis) Guyana
450 **scutella** (Zeller, 1877) (Gelechia) [S. Am?]
451 **secretella** (Walker, 1864) (Gelechia) Brazil (Am)
plejadella (R. Felder & Rogenhofer, 1875) (Gelechia) Brazil (Am)
trimolybda (Meyrick, 1914) (Anacampsis) Guyana
452 **seductella** (Walker, 1864) (Gelechia) Brazil (Am)
453 **sesamodes** Meyrick, 1922 Peru
454 **siderophaea** (Walsingham, 1910) (Anacampsis) Mexico (Ver)
455 **solidella** (Walker, 1864) (Gelechia) Brazil (Am)
456 **sporozona** (Meyrick, 1914) (Anacampsis) Guyana
457 **stasigastra** Meyrick, 1922 Brazil (Pa)
458 **stelliferella** (Walker, 1864) (Gelechia) Brazil (Am)
speciosella (Walker, 1864) (Gelechia) Brazil (Am)
459 **stillata** Meyrick, 1922 Peru
460 **subapicalis** (Walker, 1864) (Gelechia) Brazil (Am)
461 **succincta** (Walsingham, 1910) (Anacampsis) Mexico (Tab)
462 **suffectella** (Walker, 1864) (Gelechia) Brazil (Am)
463 **superfusella** (Walker, 1864) (Gelechia) Brazil (Am)
464 **susceptella** (Walker, 1864) (Gelechia) Brazil (Am)
465 **suspectella** (Walker, 1864) (Gelechia) Brazil (Am)
466 **tardella** (Walker, 1864) (Gelechia) Brazil (Am)
collocatella (Walker, 1864) (Gelechia) Brazil (Am)
sublatella (Walker, 1864) (Gelechia) Brazil (Am)
467 **terrenella** (Busck, 1914) (Anacampsis) Panama
468 **tetrortha** Meyrick, 1922 Brazil (Am)
469 **thysanora** (Meyrick, 1914) (Anacampsis) Peru
470 **titanota** (Walsingham, 1910) (Anacampsis) Guatemala
471 **trachycnemis** Meyrick, 1922 Peru
472 **trajectella** (Walker, 1864) (Gelechia) Brazil (Am)
diazeucta Meyrick, 1918, repl. name Brazil (Am)
473 **transjectella** (Walker, 1864) (Gelechia) Brazil (Am)
474 **trapezias** Meyrick, 1922 Brazil (Pa)
475 **trochilea** (Walsingham, 1910) (Anacampsis) Mexico (Tab)
476 **versatella** (Walker, 1864) (Gelechia) Brazil (Am)
477 **volubilis** Meyrick, 1922 Peru
478 **zebrina** (Walsingham, 1910) (Anacampsis) Mexico (Tab)

HOLOPHYSIS Walsingham, 1910
Hoplophysis McDunnough, 1939, missp.

479 **anoma** Walsingham, 1910 Mexico (Ver/Tab)
480 **autodesma** (Meyrick, 1918) (Zalithia) Colombia
481 **auxiliaris** (Meyrick, 1918) (Zalithia) Colombia
482 **barydesma** (Meyrick, 1918) (Zalithia) Ecuador
483 **quadrimaculata** Walsingham, 1910 Mexico (Tab)
484 **stagmatophoria** Walsingham, 1910 Mexico (Gue)
485 **tentatella** (Walker, 1864) (Gelechia) Brazil (Am)
486 **xanthostoma** Walsingham, 1910 Mexico (Gue)

STROBISIA Clemens, 1860
Systasiota Walsingham, 1910

487 **argentifrons** Walsingham, 1911 Mexico (Tab)
488 **iridipennella** Clemens, 1860 USA [Pa?]
aphroditeella Chambers, 1872 USA (Ky)
aphroditella Frey, 1878, missp.
irridipennella Busck, [1903], missp.
489 **leucura** (Walsingham, 1910) (Systasiota) Mexico (Ver)
490 **sapphiritis** Meyrick, 1914 Guyana
491 **spintheropis** Meyrick, 1922 Brazil
492 **stellaris** (R. Felder & Rogenhofer, 1875) (Pancalia) Colombia

UNTOMIA Busck, 1906

493 **acicularis** Meyrick, 1918 Ecuador
494 **alticolens** Walsingham, 1911 Mexico (Gue)
alticolans Meyrick, 1925, emend.
495 **horista** Walsingham, 1911 Mexico (Tab)
496 **juventella** (Walsingham, 1897) (Ypsolophus) Jamaica
497 **latistriga** Walsingham, 1911 Mexico (Mor)
498 **melanobathra** Meyrick, 1918 Ecuador
499 **rotundata** Walsingham, 1911 Mexico (Tab)

Chelariinae

CRASIMORPHA Meyrick, 1923

500 **infuscata** Hodges, 1963 USA (Hi)
501 **peragrapta** Meyrick, 1923 French Guiana

EPILECHIA Busck, 1939

502 **catalinella** (Busck, 1907) (Gelechia) USA (Ca)
tehuacana (Busck, 1913) (Gelechia) Mexico (DF)

HAPLOCHELA Meyrick, 1923

503 **trigonota** (Walsingham, 1911) (Psoricoptera) Mexico (Tab)
mundana (Meyrick, 1914) (Chelaria) Guyana

HYPATIMA Hübner, [1825]
Chelaria Haworth, 1828
Hypatina Stephens, 1835, missp.
Allocota Meyrick, 1904, preocc. (Motschulsky, 1860 [Coleoptera])
Cynestomorpha Meyrick, 1904
Deuteroptila Meyrick, 1904
Cymatomorpha Meyrick, 1904
Semodictis Meyrick, 1909
Allocotaniana Strand, 1913, repl. name
Episacta Turner, 1919
Cellaria Neave, 1939, missp.
Cheleria Lhomme, [1948], missp.

504 **euchorda** (Meyrick, 1923) (Chelaria) Brazil (Pa)
505 **hora** (Busck, 1914) (Psoricoptera) Panama

PECTINOPHORA Busck, 1917

506 **gossypiella** (Saunders, 1844) (Depressaria) India
umbripennis (Walsingham, 1885) (Gelechia) India

PESSOGRAPTIS Meyrick, 1923

507 **cancellata** (Meyrick, 1914) (Chelaria) Guyana
508 **cyanactis** Meyrick, 1930 Brazil (Pa)
509 **thalamias** Meyrick, 1923 Brazil (Am)

PORPODRYAS Meyrick, 1920

510 **prasinantha** Meyrick, 1920 French Guiana

PROSTOMEUS Busck, 1903

511 **brunneus** Busck, 1903 USA (Fl)

SEMOPHYLAX Meyrick, 1932

512 **apicepuncta** (Busck, 1911) (Psoricoptera) French Guiana
praesignis (Meyrick, 1913) (Anisoplaca) Peru
apicipuncta (Meyrick, 1925) (Chelaria), emend.

SITOTROGA Heinemann, 1870
Silotroga Kirby, 1871, missp.
Nesolechia Meyrick, 1921
Syngenomictis Meyrick, 1927
Sitotrogus Matsumura, 1931, missp.
Sitotrega Borg, 1932, missp.
Sititroga Costa Lima, 1945, missp.

513 **cerealella** (Olivier, 1789) (Alucita) [Europe]
hordei (Kirby, 1815) (Tinea) ?
arctella (Walker, 1864) (Gelechia) Sri Lanka
melanarthra (Lower, 1900) (Gelechia) Australia
palearis (Meyrick, 1913) (Epithectis) India
ochrescens (Meyrick, 1938) (Aristotelia) China
514 **coarctella** Zeller, 1877 Colombia

Dichomeridinae

ACOMPSIA Hübner, [1825]
Acampsia Westwood, 1840, missp.
Accompsia Bruand, 1850, missp.
Brachycrossata Heinemann, 1870
Brachicrossata Hartmann, 1880, missp.
Cathegesis Walsingham, 1910

515 **angulifera** Walsingham, 1897 Grenada
516 **psoricopterella** (Walsingham, 1892) (Brachycrossata) Grenada
517 **vinitincta** (Walsingham, 1910) (Cathegesis) Guatemala

ANORTHOSIA Clemens, 1860
Sagaritis Chambers, 1872, preocc. (Billberg, 1820 [Crustacea])
Anorthodisca Gaede, 1937, missp.

518 **capillata** Walsingham, 1911 Guatemala
519 **punctipennella** Clemens, 1860 USA [Pa?]
gracilella (Chambers, 1872) (Sagaritis) USA (Ky)

BRACHMIA Hübner, [1825]
Braclunia Stephens, 1834, missp.
Cladodes Heinemann, 1870, preocc. (Solier, 1849 [Coleoptera])
Ceratophora Heinemann, 1870, preocc. (Gray, [1835] [Reptilia])
Eudodacles Snellen, 1889, repl. name
Aulacomima Meyrick, 1904
Apethistis Meyrick, 1908

520 **chambersella** (Murtfeldt, 1874) (Gelechia) USA (Mo)
subalbusella (Chambers, 1874) (Gelechia) USA (Tx)
parvipulvella (Chambers, 1874) (Gelechia) USA (Tx)
inaequepulvella (Chambers, 1875) (Gelechia) USA (Mo)
subalbella (Walsingham, 1911) (Dichomeris), emend.
subalbella Meyrick, 1925, emend.
521 **convolvuli** (Walsingham, 1908) (Trichotaphe) Canary Is.

crypsilychna Meyrick, 1914 India
dryadopa Meyrick, 1918 South Africa
effera (Meyrick, 1918) (Lecithocera) India
emigrans (Meyrick, 1921) (Lecithocera) Barbados
522 **lyrella** (Walsingham, 1911) (Dichomeris) Guatemala
523 **virescens** Walsingham, 1911 Mexico (Gue)

BRACHYACMA Meyrick, 1886
Lathontogenus Walsingham, 1897
Paraspistes Meyrick, 1905
Lipatia Busck, 1910
Paraspistis Busck, 1914, missp.
Brachyaema Povolný, 1964, missp.
Lathontogonus Diakonoff, [1968], missp.
Brachiacma Common, 1970, missp.
Lathontogenes Hodges, 1983, missp.

524 **palpigera** (Walsingham, 1891) (Gelechia) Mozambique
adustipennis (Walsingham, 1897) (Lathontogenus) Virgin Is. (St. Croix)
ioloncha (Meyrick, 1905) (Paraspistes) Sri Lanka
crotalariella (Busck, 1910) (Lipatia) Trinidad
epichorda Turner, 1919 Australia

COTYLOSCIA Meyrick, 1923

525 **caustonota** (Meyrick, 1914) (Trichotaphe) Guyana
526 **terracocta** (Walsingham, 1911) (Anorthosia) Panama
527 **triplagella** (Walker, 1864) (Gelechia) Brazil (Am)

CYMOTRICHA Meyrick, 1923
Oxysactis Meyrick, 1923

528 **abortiva** (Walsingham, 1911) (Dichomeris) Guatemala
529 **amauropis** Meyrick, 1923 Peru
530 **amphicosma** Meyrick, 1930 Brazil (Pa)
531 **cinctella** (Walker, 1864) (Gelechia) Brazil (Am)
subrutila Meyrick, 1923 Brazil (Am)
532 **cinnamicostella** (Zeller, 1877) (Epicorthylis) Panama
533 **cyclospila** (Meyrick, 1918) (Trichotaphe) French Guiana
534 **designatella** (Walker, 1864) (Gelechia) Brazil (Am)
535 **directa** (Meyrick, 1912) (Trichotaphe) Venezuela
536 **ellipsias** (Meyrick, 1922) (Dichomeris) Peru
537 **exallacta** Meyrick, 1923 Peru
538 **fluctuans** Meyrick, 1923 Peru
539 **inspiciens** Meyrick, 1931 Paraguay
540 **intentella** (Walker, 1864) (Gelechia) Brazil (Am)
541 **jugata** (Walsingham, 1911) (Dichomeris) Mexico (Tab)
542 **melanota** (Walsingham, 1911) (Dichomeris) Mexico (Ver)
543 **melissia** (Walsingham, 1911) (Dichomeris) Panama
544 **memnonia** (Meyrick, 1913) (Trichotaphe) Brazil (RJ)
545 **miltophragma** (Meyrick, 1922) (Dichomeris) Brazil (Pa)
546 **mochlopis** Meyrick, 1923 Brazil (Am)
547 **ostensella** (Walker, 1864) (Gelechia) Brazil (Am)
548 **pectinella** (Forbes, 1931) (Trichotaphe), **n. comb.** Puerto Rico
549 **permundella** (Walker, 1864) (Gelechia) Brazil (Am)
tactella (Walker, 1864) (Gelechia) Brazil (Am)
550 **procyphodes** (Meyrick, 1922) (Dichomeris) Brazil (Am)
551 **ptilocompa** (Meyrick, 1922) (Dichomeris) Peru
552 **rubiginosella** (Walker, 1864) (Cryptolechia) Brazil (Am)
ochropyga (Walsingham, 1911) (Dichomeris) Mexico (Tab)
553 **serena** (Meyrick, 1909) (Trichotaphe) Bolivia
554 **sphyrocopa** Meyrick, 1918 French Guiana
555 **subdentata** (Meyrick, 1922) (Dichomeris) Brazil (Pa)
556 **sumptella** (Walker, 1864) (Gelechia) Brazil (Am)
557 **thalamopa** (Meyrick, 1922) (Dichomeris) Brazil (Am)
558 **thalpodes** (Meyrick, 1922) (Dichomeris) Peru
559 **themelia** (Meyrick, 1913) (Trichotaphe) Brazil (SP)
560 **thesmiopa** (Meyrick, 1922) (Dichomeris) Brazil (Pa)
561 **trigonella** (Walsingham, 1892) (Trichotaphe) St. Vincent
562 **tristicta** (Busck, 1914) (Dichomeris) Panama
563 **turrita** (Meyrick, 1914) (Trichotaphe) Guyana

DICHOMERIS Hübner, [1818]
Oxybelia Hübner, [1825]
Rhinosia Treitschke, 1833
Rhobonda Walker, 1864, preocc. (Walker, 1863 [Choreutidae])
Carna Walker, 1864, repl. name; preocc. (Gistel, 1848 [Echinodermata])
Gaeza Walker, 1864
Eurysara Turner, 1919
Euryzancla Turner, 1919
Macrozancla Turner, 1919
Brochometis Meyrick, 1923

564 **acrolychna** Meyrick, 1922 Brazil (Pa)
565 **acuminata** (Staudinger, 1876) (Mesophleps) Italy
ianthes (Meyrick, 1887) (Hypsolophus [sic]) Reunion
rusticus (Walsingham, 1892) (Ypsolophus) St. Vincent
ammoxanthus (Meyrick, 1904) (Ypsolophus) Australia
ochrophanes (Meyrick, 1907) (Ypsolophus) Sri Lanka
566 **aequata** Meyrick, 1914 Guyana
567 **antisticha** Meyrick, 1926 Costa Rica
568 **argentinella** (Berg, 1885) (Ypsolophus) Argentina
569 **baccata** Meyrick, 1923 Brazil (Am)
570 **brachymetra** Meyrick, 1923 Peru
571 **cachrydias** Meyrick, 1914 Guyana
572 **caryaefoliella** (Chambers, 1872) (Ypsolophus) USA (Ky)
caryifoliella Meyrick, 1925, emend.
573 **crossospila** Meyrick, 1933 Costa Rica
574 **dignella** Walsingham, 1911 Mexico (Gue)
575 **famulata** Meyrick, 1914 Guyana
granivora Meyrick, 1932 Trinidad
576 **fida** Meyrick, 1923 Brazil (Pa)
577 **formulata** (Meyrick, 1922) (Trichotaphe) Brazil (Am)
578 **hemichrysella** (Walker, 1863) (Psecadia) Brazil (Am)
excisorella (Walker, 1864) (Sophronia) Brazil (Am)
macroptera Meyrick, 1914 Guyana
579 **heteracma** Meyrick, 1923 Brazil (Am)
580 **hexasticta** Walsingham, 1911 Mexico (Gue)
581 **horiodes** Meyrick, 1923 Brazil (Am)
582 **horocompsa** Meyrick, 1933 Bolivia
583 **indigna** (Walsingham, 1892) (Ypsolophus) St. Vincent
584 **ingloria** Meyrick, 1923 Peru
585 **instans** Meyrick, 1923 Peru
586 **ligulella** Hübner, [1818] USA
contubernatella (Fitch, 1853) (Chaetochilus) USA (NY)
pometella (Harris, 1853) (Rhinocera) USA (Ma)
malifoliella (Fitch, 1854) (Chaetochilus) USA (NY)
flavivittella (Clemens, 1864) (Ypsolophus) USA (Ky)
quercipominella (Chambers, 1872) (Ypsolophus) USA (Ky)
reedella (Chambers, 1872) (Ypsolophus) USA (Ky)
ruderella (Chambers, 1878) (Ypsolophus), missp.
587 **lucrifuga** Meyrick, 1923 Brazil (Pa)
588 **lypetica** Walsingham, 1911 Mexico (Gue)
589 **mexicana** Walsingham, 1911 Mexico (Son)
590 **percnopholis** Walsingham, 1911 Guatemala
591 **piperata** (Walsingham, 1892) (Ypsolophus) St. Vincent
592 **plexigramma** Meyrick, 1922 Guyana
593 **prensans** Meyrick, 1922 Brazil (Pa)
594 **punctatella** (Walker, 1864) (Rhobonda) Brazil (Am)
595 **saturata** Meyrick, 1923 Brazil (Pa)
596 **sciastes** Walsingham, 1911 Mexico (Ver)
597 **squalens** Meyrick, 1914 Guyana
598 **stratella** (Walsingham, 1897) (Ypsolophus) Trinidad
599 **stratigera** Meyrick, 1922 Brazil (Am)
600 **substratella** Walsingham, 1911 Mexico (Tab)
601 **tactica** Meyrick, 1918 Ecuador
602 **thermodryas** Meyrick, 1923 Peru
603 **zomias** Meyrick, 1914 Guyana

EUNEBRISTIS Meyrick, 1923

604 **cinclidias** (Meyrick, 1918) (Noeza) French Guiana
605 **gyralea** (Meyrick, 1922) (Noeza) Brazil (Am)
606 **oncotera** (Walsingham, 1911) (Noeza) Mexico (Gue)
607 **zachroa** (Meyrick, 1914) (Noeza) Guyana
608 **zingarella** (Walsingham, 1897) (Malacotricha) Virgin Is. (St. Croix)

HAPALONOMA Meyrick, 1914

609 **argyracta** Meyrick, 1914 Guyana
610 **sublustricella** (Walker, 1864) (Gelechia) Brazil (Am)

ILINGIOTIS Meyrick, 1914
Sirogenes Meyrick, 1923

611 **hemeropa** Meyrick, 1923 Brazil (Am)
612 **sevectella** (Walker, 1864) (Gelechia) Brazil (Am)
613 **thermophaea** (Meyrick, 1923) (Sirogenes) Brazil (Am)
614 **thrasynta** Meyrick, 1914 Guyana
615 **vigilans** Meyrick, 1914 Guyana

MYROPHILA Meyrick, 1923

616 **carycina** (Meyrick, 1914) (Trichotaphe) Guyana
617 **caryophragma** Meyrick, 1923 Guyana
618 **diacnista** Meyrick, 1923 Guyana

ONEBALA Walker, 1864
Helcystogramma Zeller, 1877
Dectobathra Meyrick, 1904

619 **adaequata** (Meyrick, 1914) (Helcystogramma) Guyana
adequata Clarke, 1969, missp.
620 **anisopa** (Meyrick, 1918) (Anacampsis) Colombia
621 **archigrapha** Meyrick, 1929 Colombia
622 **carycastis** (Meyrick, 1922) (Helcystogramma) Brazil (Pa)
623 **cerinura** (Meyrick, 1923) (Brachmia) Brazil (Pa)
624 **chalyburga** (Meyrick, 1922) (Helcystogramma) Brazil (Pa)
625 **daedalea** (Walsingham, 1911) (Dichomeris) Mexico (Tab)
626 **elliptica** (Forbes, 1931) (Trichotaphe), **n. comb.** Puerto Rico
627 **meconitis** (Meyrick, 1913) (Trichotaphe) Argentina
628 **ribeella** (Zeller, 1877) (Helcystogramma) Panama
629 **rusticella** (Walker, 1864) (Gelechia) Brazil (Am)
630 **sertigera** (Meyrick, 1923) (Helcystogramma) Peru
631 **stellatella** Busck, 1914 (Dichomeris) Panama

632 **symbolica** (Meyrick, 1914) (Helcystogramma) Guyana
633 **tegulella** (Walsingham, 1897) (Trichotaphe) Grenada
servilis (Walsingham, 1911) (Dichomeris) Panama
634 **trichocyma** (Meyrick, 1923) (Brachmia) Brazil (Am)

PACHYSARIS Meyrick, 1914

635 **collina** Meyrick, 1914 Peru
636 **contrita** Meyrick, 1922 Brazil (Pa)
637 **paenitens** Meyrick, 1923 Brazil (Am)
638 **rurigena** Meyrick, 1914 Guyana

PARANOEA Walsingham, 1911

639 **latescens** Walsingham, 1911 Mexico (Tab)

PLOCAMOSARIS Meyrick, 1912
Noeza Walker, 1866, preocc. (Meigen, 1800 [Diptera])
Neochrista Meyrick, 1923

640 **auritogata** (Walsingham, 1911) (Noeza) Panama
641 **pandora** Meyrick, 1912 Brazil (RJ)
642 **pyropis** (Meyrick, 1918) (Noeza) French Guiana
643 **telegraphella** (Walker, 1866) (Noeza) Brazil (Am)

PROPHORAULA Meyrick, 1922

644 **pyrrhopis** Meyrick, 1922 Brazil (Am)

SEMIOMERIS Meyrick, 1923

645 **pyretodes** Meyrick, 1914 Guyana

TEUCHOPHANES Meyrick, 1914

646 **cornuta** (Busck, 1914) (Dichomeris) Panama
647 **luminosa** (Busck, 1914) (Dichomeris) Panama
leucopleura Meyrick, 1914 Guyana
648 **perceptella** (Busck, 1914) (Dichomeris) Panama

TRICHOTAPHE Clemens, 1860
Begoe Chambers, 1862
Malacotricha Zeller, 1873
Tricotaphe Riley, 1891, missp.
Malacotriche Busck, 1903, missp.
Malachotriche Busck, 1903, missp.

649 **aenigmatica** Clarke, 1962 Mexico (Ver)
650 **anticrates** Meyrick, 1931 Brazil
651 **ardesiella** (Walsingham, 1911) (Dichomeris) Mexico (Ver)
652 **argigastra** (Walsingham, 1911) (Dichomeris) Mexico (Ver)
argogastra Meyrick, 1925, missp.
653 **arotrosema** (Walsingham, 1911) (Dichomeris) Mexico (Ver)
654 **atriguttata** Meyrick, 1931 Paraguay
655 **aurisulcata** Meyrick, 1922 Brazil (Am)
656 **carinella** (Walsingham, 1911) (Dichomeris) Mexico (Gue)
657 **chalinopis** Meyrick, 1935 Argentina
*658 **condaliavorella** Busck, 1900, mispl. USA (Fl)
659 **cyanoneura** Meyrick, 1922 Brazil (Pa)
660 **euparypha** Meyrick, 1922 Peru
661 **evitata** (Walsingham, 1911) (Dichomeris) Panama
662 **excavata** (Busck, 1914) (Dichomeris) Panama
663 **fulvicilia** Meyrick, 1922 Brazil (Am)
664 **habrochitona** (Walsingham, 1911) (Dichomeris) Panama
665 **manella** (Möschler, 1890) (Ypsolophus) Puerto Rico
666 **mistipalpis** (Walsingham, 1911) (Dichomeris) Panama
violaria Meyrick, 1914 Guyana
667 **nessica** (Walsingham, 1911) (Dichomeris) Panama
668 **opsonoma** (Meyrick, 1914) (Dichomeris) Guyana
669 **porphyrogramma** Meyrick, 1914 Guyana
670 **renascens** (Walsingham, 1911) (Dichomeris) Mexico (Tab)
671 **retracta** Meyrick, 1922 Brazil (Pa)
672 **semicuprata** Meyrick, 1922 Peru
673 **syringota** Meyrick, 1926 Peru
674 **tangolias** Gyen, 1913 Chile
675 **thrypsandra** Meyrick, 1923 Ecuador
676 **varronia** (Busck, 1913) (Dichomeris) Guyana
677 **vetustella** (Walker, 1864) (Paepia) Brazil (Am)
678 **xerodes** (Walsingham, 1911) (Dichomeris) Mexico (Tab)
679 **xuthostola** (Walsingham, 1911) (Dichomeris) Mexico (Tab)

VAZUGADA Walker, 1864

680 **abscessella** (Walker, 1863) (Psecadia) Brazil (Am)
strigiplenella Walker, 1864 Brazil (Am)
zonostoma (Meyrick, 1914) (Dichomeris) Guyana
681 **amphicoma** (Meyrick, 1912) (Dichomeris) Brazil (SP)
682 **costalis** (Busck, 1914) (Dichomeris) Panama
683 **leucostena** (Walsingham, 1911) (Dichomeris) Mexico (Gue)
684 **macrosphena** (Meyrick, 1913) (Trichotaphe) Brazil (SP)
685 **percnacma** Meyrick, 1923 Peru
686 **phoenogramma** Meyrick, 1930 Brazil (Pa)

Lecithocerinae

DEOCLONA Busck, 1903
Proclesis Walsingham, 1911
Lioclepta Meyrick, 1922
Deoclana Fletcher, 1929, missp.

687 **complanata** (Meyrick, 1922) (Lioclepta), **n. comb.** Peru
688 **eriobotryae** Busck, 1939 Argentina
689 **xanthoselene** (Walsingham, 1911) (Proclesis) Guyana
xanthoselena Meyrick, 1925, emend.

Unplaced Genera

ACROPHILETES Meyrick, 1932

690 **cosmocrossa** Meyrick, 1932 Bolivia

ADULLAMITIS Meyrick, 1932
Adullanitis Gaede, 1937, missp.

691 **emancipata** Meyrick, 1932 Brazil (Pa)

AEROTYPIA Walsingham, 1911

692 **pleurotella** Walsingham, 1911 Mexico (Mor)

AGATHACTIS Meyrick, 1929

693 **toxocosma** Meyrick, 1929 Guyana

AGELIARCHES Meyrick, 1923

694 **rhizogramma** Meyrick, 1923 Brazil (Pa)

ALSODRYAS Meyrick, 1914

695 **deltochlora** Meyrick, 1922 Brazil (Pa)
696 **lactaria** Meyrick, 1914 Guyana
697 **prasinoptila** Meyrick, 1922 Brazil (Pa)

ANOMOXENA Meyrick, 1917

698 **spinigera** Meyrick, 1917 Colombia
699 **tetraxoa** Meyrick, 1917 Ecuador

ANTHINORA Meyrick, 1914

700 **xanthophanes** Meyrick, 1914 Guyana

ANTHISTARCHA Meyrick, 1925
Antistarcha Costa Lima, 1945, missp.

701 **binocularis** Meyrick, 1929 Brazil (Ba)
702 **geniatella** (Busck, 1914) (Gelechia) Panama

APOTACTIS Meyrick, 1918

703 **citroptila** Meyrick, 1933 Costa Rica

APOTHETOECA Meyrick, 1922

704 **synaphrista** Meyrick, 1922 Chile (Juan Fernandez Is.)

AROTROMIMA Meyrick, 1929

705 **politica** Meyrick, 1929 Guyana

BELTHECA Busck, 1914
Anterethista Meyrick, 1914
Antherethista Gaede, 1937, missp.

706 **phosphoropa** (Meyrick, 1922) (Anterethista) Peru
707 **picolella** Busck, 1914 Panama
heteractis (Meyrick, 1914) (Anterethista) Guyana

BESCIVA Busck, 1914

708 **longitudinella** Busck, 1914 Panama

BRACHYPSALTIS Meyrick, 1931

709 **subalbata** Meyrick, 1931 Argentina

BRUCHIANA Jörgensen, 1916

710 **cassiaella** Jörgensen, 1916 Argentina

CATALEXIS Walsingham, 1909

711 **tapinota** Walsingham, 1909 Guatemala

CATOPTRISTIS Meyrick, 1925

712 **trissoxantha** (Meyrick, 1922) (Strobisia) Peru

CERYCANGELA Meyrick, 1925

713 **sacricola** (Meyrick, 1922) (Zalithia) Brazil (Am)

CHALCOMIMA Meyrick, 1929

714 **hoplodoxa** Meyrick, 1929 Peru

CHARISTICA Meyrick, 1925

715 **caeligena** (Meyrick, 1922) (Zalithia) Brazil (Am)
716 **callichroma** (Meyrick, 1914) (Zalithia) Guyana
717 **exteriorella** (Walker, 1864) (Gelechia) Brazil (Am)
718 **ioploca** (Meyrick, 1922) (Zalithia) Brazil (Am)
719 **iriantha** (Meyrick, 1914) (Zalithia) Guyana
720 **porphyraspis** (Meyrick, 1909) (Strobisia) Bolivia
721 **rhodopetala** (Meyrick, 1922) (Zalithia) Brazil (Am)
722 **sandaracota** (Meyrick, 1914) (Zalithia) Guyana
723 **walkeri** (Walsingham, 1911) (Strobisia) Panama
euphracta (Meyrick, 1914) (Zalithia) Guyana

CLISTOTHYRIS Zeller, 1877

724 **villosula** Zeller, 1877 Colombia

COLEOSTOMA Meyrick, 1922

725 **entryphopa** Meyrick, 1922 Brazil (Pa)

COLONANTHES Meyrick, 1923

726 **plectanopa** Meyrick, 1923 Peru

COMMATICA Meyrick, 1909
Apopira Walsingham, 1911

727 **acropelta** Meyrick, 1914 Guyana
728 **bifuscella** (Forbes, 1931) (Anacampsis) Puerto Rico
729 **chionura** Meyrick, 1914 Guyana
730 **crossotorna** Meyrick, 1929 Guyana
731 **cryptina** (Walsingham, 1911) (Untomia) Mexico (Tab)
732 **cyanorrhoa** Meyrick, 1914 Guyana
733 **emplasta** Meyrick, 1914 Guyana
734 **eremna** Meyrick, 1909 Bolivia
735 **extremella** (Walker, 1864) (Gelechia) Brazil (Am)
736 **falcatella** (Walker, 1864) (Gelechia) Brazil (Am)
rostella (R. Felder & Rogenhofer, 1875) (Gelechia) Brazil (RJ)
737 **hexacentra** Meyrick, 1922 Brazil (Am)
738 **lupata** Meyrick, 1914 Guyana
739 **metochra** Meyrick, 1914 Guyana
740 **nerterodes** Meyrick, 1914 Guyana
741 **palirrhoa** Meyrick, 1922 Brazil (Am)
742 **parmulata** Meyrick, 1914 Guyana
743 **phanocrossa** Meyrick, 1922 Brazil (Am)
744 **placoterma** Meyrick, 1918 Colombia
745 **pterygota** Meyrick, 1929 Brazil (Pa)
746 **servula** Meyrick, 1922 Peru
747 **stygia** Meyrick, 1922 Brazil (Am)
748 **xanthocarpa** Meyrick, 1922 Peru

COMPSOSARIS Meyrick, 1914
Gompsosaris Gaede, 1937, missp.

749 **flavidella** (Busck, 1914) (Recurvaria) Panama
750 **testacea** Meyrick, 1914 Guyana

COPTICOSTOLA Meyrick, 1929

751 **acuminata** (Walsingham, 1911) (Untomia) Mexico (Tab)

CRAMBODOXA Meyrick, 1913 [29]

752 **platyaula** Meyrick, 1913 Colombia

DARLIA Clarke, 1950

753 **praetexta** Clarke, 1950 Argentina

DIASTALTICA Walsingham, 1910

754 **separabilis** Walsingham, 1910 Guatemala

DISSOPTILA Meyrick, 1914

755 **asphaltitis** Meyrick, 1914 Guyana
756 **crocodora** Meyrick, 1922 Peru
757 **disrupta** Meyrick, 1914 Guyana
758 **mutabilis** Meyrick, 1914 Guyana
759 **prozona** Meyrick, 1914 Guyana

DREPANOTERMA Walsingham, 1897

760 **lacticaudellum** Walsingham, 1897 Grenada

ELASIPRORA Meyrick, 1914

761 **rostrigera** Meyrick, 1914 Guyana

EMPEDAULA Meyrick, 1918

762 **phanerozona** Meyrick, 1922 Brazil (Pa)
763 **rhodocosma** (Meyrick, 1914) (Aristotelia) Guyana

ERIPNURA Meyrick, 1914

764 **criodes** Meyrick, 1914 Guyana

ERISTHENODES Meyrick, 1935

765 **tetrapetra** Meyrick, 1935 Argentina

ERYTHRIASTIS Meyrick, 1925

766 **rhodocrossa** (Meyrick, 1914) (Pachnistis) Guyana
767 **rubentula** (Meyrick, 1914) (Pachnistis) Guyana

ETHIROSTOMA Meyrick, 1914

768 **interpolata** Meyrick, 1922 Peru
769 **semiacma** Meyrick, 1914 Guyana

EUNOMARCHA Meyrick, 1923
Atoponeura Busck, 1914, preocc. (Szépligeti, 1905 [Hymenoptera])

770 **violacea** (Busck, 1914) (Atoponeura) Panama
glycinopis Meyrick, 1923 Brazil (Am)

EUZONOMACHA Meyrick, 1925

771 **subjectella** (Walker, 1864) (Gelechia) Brazil (Am)

FORTINEA Busck, 1914

772 **auriciliella** Busck, 1914 Panama

GALTICA Busck, 1914

773 **venosa** Busck, 1914 Panama

GLAUCACNA Forbes, 1931
Glaucagna Gaede, 1937, missp.

774 **iridea** Forbes, 1931 Puerto Rico

ILARCHES Meyrick, 1933

775 **notaula** Meyrick, 1933 Bolivia

IPHIMACHAERA Meyrick, 1931 [30]

776 **picticollis** (Walsingham, 1912) (Mnesichara) Panama
decapitata Meyrick, 1931 Brazil (SP)

ISEMBOLA Meyrick, 1926

777 **diasticta** Meyrick, 1926 Ecuador

LEISTOGENES Meyrick, 1927 [31]

778 **rebellis** Meyrick, 1927 Peru

LOGISIS Walsingham, 1909

779 **achroea** Walsingham, 1909 Costa Rica

LOPHAEOLA Meyrick, 1932

780 **inquinata** Meyrick, 1932 Brazil (Am)

MERIDORMA Meyrick, 1925

781 **thrombodes** (Meyrick, 1914) (Ptocheuusa) Guyana

METABOLAEA Meyrick, 1923

782 **chlorophthalma** Meyrick, 1923 Brazil (Am)

METOPLEURA Busck, 1912

783 **potosi** Busck, 1912 Mexico (SLP)

MOLOPOSTOLA Meyrick, 1920

784 **calumnians** Meyrick, 1926 Brazil (Pa)
785 **rufitecta** Meyrick, 1920 French Guiana

OESTOMORPHA Walsingham, 1911

786 **alloea** Walsingham, 1911 Mexico (Ver)

OXYCRYPTIS Meyrick, 1912

787 **attonita** Meyrick, 1912 Colombia

OXYLECHIA Meyrick, 1917

788 **confirmata** Meyrick, 1917 Colombia

PACHYGENEIA Meyrick, 1923

789 **clitellaria** Meyrick, 1923 Brazil (Am)
oxyloba (Meyrick, 1929) (Gnorimoschema) Brazil (Am)

PARASTEGA Meyrick, 1912

790 **chionostigma** (Walsingham, 1911) (Psoricoptera) Panama
ochropis Meyrick, 1914 Ecuador/Surinam
791 **hemisigna** Clarke, 1951 Argentina
792 **niveisignella** (Zeller, 1877) (Psoricoptera) Panama
curvatella Busck, 1914 Panama
793 **trichella** Busck, 1914 Panama

PARELECTROIDES Clarke, 1952, repl. name
Parelectra Meyrick, 1925, preocc. (Dognin, 1914 [Noctuidae])

794 **chalybea** (R. Felder & Rogenhofer, 1875) (Simaethis), **n. comb.** Brazil (Am)
795 **helicopis** (Meyrick, 1922) (Strobisia) Brazil (Pa)
796 **scintillula** (Walsingham, 1911) (Strobisia) Mexico (Tab)
797 **selectella** (Walker, 1864) (Gelechia) Brazil (Am)
798 **subvectella** (Walker, 1864) (Gelechia) Brazil (Am)

PAVOLECHIA Busck, 1914
Desmaucha Meyrick, 1918

799 **argentea** Busck, 1914 Panama
chrysostoma (Meyrick, 1918) (Desmaucha) Guyana

PELOCNISTIS Meyrick, 1932

800 **xylozona** Meyrick, 1932 Brazil (Am)

PERIORISTICA Walsingham, 1910

801 **chalcopera** Walsingham, 1910 Mexico (Gue)

PHYLOPATRIS Meyrick, 1923

802 **terpnodes** Meyrick, 1923 Brazil (Pa)

PROMOLOPICA Meyrick, 1925

803 **epiphantha** Meyrick, 1925 Brazil (Pa)

PTILOSTONYCHIA Walsingham, 1911 [32]
Ptilonostychia Fletcher, 1929, missp.

804 **plicata** Walsingham, 1911 Panama

RHYNCHOTONA Meyrick, 1923

805 **phaeostrota** Meyrick, 1923 Peru

SATRAPODOXA Meyrick, 1925

806 **regia** (Meyrick, 1914) (Strobisia) Guyana

SCLEROGRAPTIS Meyrick, 1923

807 **oxytypa** Meyrick, 1923 Guyana

SIMONEURA Walsingham, 1911

808 **ophitis** Walsingham, 1911 Mexico (Tab)

SOROTACTA Meyrick, 1914

809 **bryochlora** Meyrick, 1922 Brazil (Am)
810 **viridans** Meyrick, 1914 Guyana

STACHYOSTOMA Meyrick, 1923

811 **psilodoxa** Meyrick, 1923 Ecuador

STAGMATURGIS Meyrick, 1923

812 **catharosema** Meyrick, 1923 Brazil (Am)

STEREMNIODES Meyrick, 1923

813 **sciactis** Meyrick, 1923 Brazil (Am)

STEREODMETA Meyrick, 1931

814 **xylodeta** Meyrick, 1931 Brazil (ES)

STIBARENCHES Meyrick, 1930

815 **bifissa** Meyrick, 1930 Brazil (Pa)

SYMPHANACTIS Meyrick, 1925

816 **hetaera** (Meyrick, 1914) (Ptocheuusa) Guyana

SYNACTIAS Meyrick, 1931

817 **micranthis** Meyrick, 1931 Paraguay

TABERNILLAIA Walsingham, 1911
Tabernillaea Meyrick, 1925, emend.

818 **ephialtes** Walsingham, 1911 Panama

TAPHROSARIS Meyrick, 1922

819 **malthacopa** Meyrick, 1922 Brazil (Am)

TECIA Kieffer & Jörgensen, 1910
Fapua Kieffer & Jörgensen, 1910
Lata Kieffer & Jörgensen, 1910

820 **albinervella** (Kieffer & Jörgensen, 1910) (Fapua) Argentina
821 **kiefferi** Kieffer & Jörgensen, 1910 Argentina
822 **mendozella** Kieffer & Jörgensen, 1910 Argentina

THRYPSIGENES Meyrick, 1914
Thripsigenes Clarke, 1955, missp.

823 **colluta** Meyrick, 1914 Guyana
824 **furvescens** Meyrick, 1914 Guyana

THYRSOMNESTIS Meyrick, 1929

825 **ceramoxantha** Meyrick, 1929 Colombia

TOCMIA Walker, 1864

826 **versicolorella** Walker, 1864 Brazil (Am)

TRICHEMBOLA Meyrick, 1918

827 **idiarcha** Meyrick, 1931 Paraguay

ZELOSYNE Walsingham, 1911

828 **olga** Meyrick, 1915 Guyana
829 **poecilosoma** Walsingham, 1911 Panama

Unplaced Species

830 "Gaea" **lilloi** Köhler, 1941, mispl. [33] Argentina

Copromorphoidea [34]

30. COPROMORPHIDAE

by J. B. Heppner

CATHELOTIS Meyrick, 1926

1 **sanidopa** Meyrick, 1926 Colombia

ENDOTHAMNA Meyrick, 1922

2 **marmarocyma** Meyrick, 1922 Chile

LOTISMA Busck, 1909

3 **trigonana** (Walsingham, 1879) (Sciaphila) USA (Ca)
kincaidiella (Busck, 1904) (Hemerophila) USA (Wa)
4 **vulcanicola** Meyrick, 1932 Costa Rica

NEOPHYLARCHA Meyrick, 1926

5 **helicosema** Meyrick, 1926 Guyana/French Guiana

ORDRUPIA Busck, 1911
Ordupia Busck, 1911, missp.
Ardrupia Busck, 1911, missp.

6 **fanniella** Busck, 1912 Panama
7 **friserella** Busck, 1911 French Guiana
fabricata Meyrick, 1915 Guyana
dasyleuca Meyrick, 1926 Peru
8 **macroctenis** Meyrick, 1926 Peru

PHYCOMORPHA Meyrick, 1914

9 **escharitis** Meyrick, 1916 Colombia

RHOPALOSETIA Meyrick, 1926

10 **phlyctaenopa** Meyrick, 1926 French Guiana

SARIDACMA Meyrick, 1930

11 **ilyopis** Meyrick, 1930 Brazil (Pa)

SYNCAMARIS Meyrick, 1932

12 **argophthalma** Meyrick, 1932 Brazil (SC)

31. ALUCITIDAE

by S. E. Miller

ALINGUATA Fleming, 1948

1 **neblina** Fleming, 1948 Venezuela

ALUCITA Linnaeus, 1758
Orneodes Latreille, 1796, suppr. (ICZN Op. 450)
Euchiradia Hübner, [1825]
Alucitina Heydenreich, 1851

2 **acalles** (Walsingham, 1915) (Orneodes), **n. comb.** Costa Rica
3 **ancalopa** (Meyrick, 1921) (Orneodes), **n. comb.** Brazil/Guyana
4 **arriguttii** (Pastrana, 1960) (Orneodes), **n. comb.** Bolivia
5 **bridarollii** (Pastrana, 1951) (Orneodes), **n. comb.** Paraguay
6 **brunnea** (Fletcher, 1926) (Orneodes), **n. comb.** Brazil (Am)
7 **eudactyla** R. Felder & Rogenhofer, 1875 Colombia/Brazil
8 **flavicincta** (Walsingham, 1915) (Orneodes), **n. comb.** Jamaica
9 **jujuyensis** (Pastrana, 1953) (Orneodes), **n. comb.** Argentina
10 **mulciber** (Meyrick, 1932) (Orneodes), **n. comb.** Costa Rica
11 **nasuta** Zeller, 1877 Colombia
12 **nubifera** (Meyrick, 1921) (Orneodes), **n. comb.** Colombia
13 **panolbia** (Walsingham, 1915) (Orneodes), **n. comb.** Guatemala
14 **patria** (Meyrick, 1921) (Orneodes), **n. comb.** Guyana
15 **proseni** (Pastrana, 1951) (Orneodes), **n. comb.** Argentina
16 **punctiferella** Walker, 1866 Belize
17 **riggii** (Orfila, 1949) (Orneodes), **n. comb.** Argentina
18 **sertifera** (Meyrick, 1921) (Orneodes), **n. comb.** French Guiana
19 **stephanopsis** (Meyrick, 1921) (Orneodes), **n. comb.** Brazil (Am)
20 **tandilensis** (Pastrana, 1960) (Orneodes), **n. comb.** Argentina
21 **trachydesma** (Meyrick, 1929) (Orneodes), **n. comb.** Bolivia

HEXERETMIS Meyrick, 1929

22 **argo** Meyrick, 1929 Peru
23 **willineri** Pastrana, 1953 Bolivia

PAELIA Walker, 1866

24 **lunuligera** Walker, 1866 Brazil (Am)

32. CARPOSINIDAE

by J. B. Heppner

CARPOSINA Herrich-Schäffer, 1853
Subgenus **Carposina** Herrich-Schäffer, 1853

1 **engalactis** Meyrick, 1932 Brazil (SC)
2 **phycitana** Walsingham, 1914 Panama

Subgenus **Trepsitypa** Meyrick, 1913

3 **cardinata** (Meyrick, 1911) (Trepsitypa) Guyana

Subgenus **Dipremna** Davis, 1969

4 **cretata** Davis, 1969 Puerto Rico

Subgenus **Epipremna** Davis, 1969

5 **dominicae** Davis, 1969 Dominica

Subgenus **Hypopremna** Davis, 1969

6 **bullata** Meyrick, 1913 Guyana

ATOPOSEA Davis, 1969

7 **maxima** (Meyrick, 1912) (Carposina) Colombia

33. EPERMENIIDAE

by J. B. Heppner

Ochromolopinae

PAROCHROMOLOPIS Gaedike, 1977

1 **parishi** Gaedike, 1977 Peru
2 **psittacanthus** Heppner, 1980 Costa Rica
3 **syncrata** (Meyrick, 1921) (Epermenia) Peru

34. GLYPHIPTERIGIDAE

by J. B. Heppner

COTAENA Walker, [1865]

1 **mediana** Walker, [1865] Brazil (Pa)
2 **plenella** (Busck, 1914) (Glyphipteryx [sic]) Panama

MYRSILA Boisduval, [1875]

3 **auripennis** Boisduval, [1875] Brazil (Pa)

PHALERARCHA Meyrick, 1913

4 **chrysorma** Meyrick, 1913 Guyana
5 **eumitrella** (Busck, 1914) (Ussara) Panama

CRONICOMBRA Meyrick, 1920

6 **deltodes** (Walsingham, 1914) (Glyphipteryx [sic]) Mexico (Tab)
7 **essedaria** Meyrick, 1926 Peru
8 **granulata** Meyrick, 1920 Brazil (Pa)
9 **lamella** (Busck, 1914) (Porpe) Panama
10 **palpella** (Walsingham, 1914) (Glyphipteryx [sic]) Mexico (Tab)
11 **phaeobathra** (Meyrick, 1932) (Ussara) Brazil (Go)
12 **porphyrospila** (Meyrick, 1926) (Machlotica) Peru

TAENIOSTOLELLA Fletcher, 1940, repl. name
Taeniostola Meyrick, 1920, preocc. (Bezzi, 1913 [Diptera])

13 **celophora** (Meyrick, 1920) (Taeniostola) Brazil (Pa)
14 **litura** (R. Felder & Rogenhofer, 1875) (Oecophora) Brazil (Am)

MACHLOTICA Meyrick, 1909
Maclotica Busck, 1915, missp.

15 **chrysodeta** Meyrick, 1909 Bolivia
16 **eurymolybda** Meyrick, 1926 Peru

NEOMACHLOTICA Heppner, 1981

17 **actinota** (Walsingham, 1914) (Glyphipteryx [sic]) Mexico (Tab)
18 **atractias** (Meyrick, 1909) (Machlotica) Bolivia
19 **nebras** (Meyrick, 1909) (Machlotica) Bolivia

TRAPEZIOPHORA Walsingham, 1892

20 **gemmula** Walsingham, 1892 St. Vincent

RHABDOCRATES Meyrick, 1931

21 **sporomantis** Meyrick, 1931 Peru

USSARA Walker, 1864
Setiostoma R. Felder & Rogenhofer, 1875 (not Zeller, 1875 [Oecophoridae])
Usara Busck, [1934], missp.

22 **ancobathra** Meyrick, 1932 Brazil (Am)
23 **ancyristis** Meyrick, 1920 Brazil (Pa)
24 **arquata** Meyrick, 1926 Colombia
25 **chalcodesma** Meyrick, 1913 Guyana
26 **chrysangela** Meyrick, 1922 Peru
27 **decoratella** Walker, 1864 Brazil (Am)
28 **eurythmiella** Busck, 1914 Panama
29 **flaviceps** (R. Felder & Rogenhofer, 1875) (Setiostoma) Brazil (Am)
30 **olyranta** Meyrick, 1931 Brazil (RS)

SERICOSTOLA Meyrick, 1927[35]

31 **rhodanopa** Meyrick, 1927 Colombia

GLYPHIPTERIX Hübner, [1825]
Heribeia Stephens, 1829
Aechmia Treitschke, 1833
Aecimia Boisduval, 1836, missp.
Glyphipteryx Zeller, 1839, emend. (not Curtis, 1827 [Cosmopterigidae])
Glyphiteryx Fischer von Röslerstamm, 1841, missp.
Anacampsoides Bruand, 1850, nom. oblit.
Glypipteryx Stainton, 1854, missp.
Glyphopteryx Herrich-Schäffer, 1854, emend.
Glyphiptoryx Mann & Rogenhofer, 1878, missp.
Glyphptieryx Turati, 1879, missp.
Glyphipterys Christoph, 1882, missp.
Glyphyteryx Hampson, 1918, missp.
Glyphteryx Watt, 1920, missp.

32 **atelura** Meyrick, 1920 (Glyphipteryx [sic]) Guyana
33 **callidelta** Meyrick, 1922 (Glyphipteryx [sic]) Peru
34 **caudatella** Walsingham, 1897 (Glyphipteryx [sic]) Grenada
35 **cestrota** Meyrick, 1915 (Glyphipteryx [sic]) Peru
36 **chrysallacta** Meyrick, 1922 (Glyphipteryx [sic]) Peru
37 **colorata** Meyrick, 1913 (Glyphipteryx [sic]) Guyana
38 **columnaris** Meyrick, 1913 (Glyphipteryx [sic]) Guyana
39 **conosema** Meyrick, 1913 (Glyphipteryx [sic]) Guyana
40 **cornigerella** Zeller, 1877 (Glyphipteryx [sic]) Colombia
41 **crinita** Meyrick, 1913 (Glyphipteryx [sic]) Guyana
42 **epastra** Meyrick, 1922 (Glyphipteryx [sic]) Peru
43 **expurgata** Meyrick, 1922 (Glyphipteryx [sic]) Peru
44 **falcigera** Meyrick, 1913 (Glyphipteryx [sic]) Guyana
45 **hologramma** Meyrick, 1920 (Glyphipteryx [sic]) Brazil (Pa)
46 **indomita** Meyrick, 1922 (Glyphipteryx [sic]) Brazil (Am)
47 **invicta** Meyrick, 1920 (Glyphipteryx [sic]) Brazil (Pa)
iinvicta Clarke, 1969, missp.
48 **ioclista** Meyrick, 1913 (Glyphipteryx [sic]) Guyana
49 **leptocona** Meyrick, 1922 (Glyphipteryx [sic]) Peru
50 **neochorda** Meyrick, 1922 (Glyphipteryx [sic]) Peru
51 **nugella** R. Felder & Rogenhofer, 1875 (Glyphipteryx [sic]) Colombia
52 **oligastra** Meyrick, 1926 (Glyphipteryx [sic]) Colombia
53 **orthodeta** Meyrick, 1922 (Glyphipteryx [sic]) Brazil (Am)
54 **oxyglypta** (Meyrick, 1929) (Acrolepia) Panama
55 **paradisea** Walsingham, 1897 (Glyphipteryx [sic]) Grenada
56 **perfracta** Meyrick, 1922 (Glyphipteryx [sic]) Peru
57 **platyochra** Meyrick, 1920 (Glyphipteryx [sic]) Brazil (Pa)
58 **pseudostoma** Meyrick, 1922 (Glyphipteryx [sic]) Guyana
59 **refractella** Zeller, 1877 (Glyphipteryx [sic]) Colombia
60 **repletana** (Walker, 1864) (Ussara) Brazil (Am)
61 **septemstrigella** Zeller, 1877 (Glyphipteryx [sic]) Colombia
62 **speculans** Meyrick, 1922 (Glyphipteryx [sic]) Brazil (Am)
63 **stasichlora** Meyrick, 1931 (Glyphipteryx [sic]) Peru
64 **syndecta** Meyrick, 1915 (Glyphipteryx [sic]) Peru
65 **synorista** Meyrick, 1922 (Glyphipteryx [sic]) Brazil (Am)
66 **uncta** Meyrick, 1913 (Glyphipteryx [sic]) Guyana
67 **unguifera** Meyrick, 1922 (Glyphipteryx [sic]) Peru
68 **variata** Meyrick, 1913 (Glyphipteryx [sic]) Guyana
69 **versicolor** Meyrick, 1913 (Glyphipteryx [sic]) Guyana
70 **voluptella** R. Felder & Rogenhofer, 1875 (Glyphipteryx [sic]) Brazil (Am)
71 **xanthoplecta** Meyrick, 1922 (Glyphipteryx [sic]) Peru
72 **zalodisca** Meyrick, 1920 (Glyphipteryx [sic]) Brazil (Pa)

DIPLOSCHIZIA Heppner, 1981

73 **glaucophanes** (Meyrick, 1922) (Glyphipteryx [sic]) Brazil (Am)
74 **tetratoma** (Meyrick, 1913) (Glyphipteryx [sic]) Guyana
75 **urophora** (Walsingham, 1914) (Glyphipteryx [sic]) Guatemala

Yponomeutoidea

35. PLUTELLIDAE

by J. B. Heppner

Plutellinae

CALLIATHLA Meyrick, 1931

1 **peplophanes** Meyrick, 1931 Argentina

EUCALLIATHLA Clarke, 1967

2 **candidella** (Blanchard, 1852) (Aecophora [sic]) Chile

EUCERATIA Walsingham, 1881

3 **argentea** (Busck, 1912) (Calantica) Mexico (Ver)

EUDOLICHURA Clarke, 1965

4 **exuta** Clarke, 1965 Chile

LEUROPERNA Clarke, 1965

5 **leioptera** Clarke, 1965 Chile (Juan Fernandez Is.)

ORTHENCHES Meyrick, 1886

6 **clytandra** Meyrick, 1931 Chile
7 **dissimulatrix** Meyrick, 1931 Chile
8 **osteacma** Meyrick, 1931 Argentina
9 **semicretata** Meyrick, 1931 Argentina

PHILAUSTERA Meyrick, 1927

10 **signifera** Meyrick, 1927 Colombia

PLUTELLA Schrank, 1802
Anadetia Hübner, [1825]
Euota Hübner, [1825]
Cerostoma.- Stephens, 1834 (not Latreille, 1802)
Creagria Sodoffsky, 1837

11 **acrodelta** Meyrick, 1931 Argentina
12 **culminata** Meyrick, 1931 Argentina
13 **deltodoma** Meyrick, 1931 Chile
14 **diluta** Meyrick, 1931 Chile
15 **nephelaegis** Meyrick, 1931 Argentina
16 **porrectella** (Linnaeus, 1758) (Phalaena Tinea) [Sweden?]
hesperidella (Hübner, 1796) (Tinea) [Europe]
vigilaciella Clemens, 1860 USA [Pa?]
17 **rectivittella** Zeller, 1877 Colombia
18 **xylostella** (Linnaeus, 1758) (Phalaena Tinea) [Sweden?]
cinerea (Geoffroy, 1785) (Tinea) [Europe]
maculipennis (Curtis, 1832) (Cerostoma) UK (England)
annulatellus Wood, 1839 UK (England)
cruciferarum Zeller, 1843 Germany
brassicella Fitch, 1856 USA (NY)
limbipennella Clemens, 1860 USA [Pa?]
mollipedella Clemens, 1860 USA [Pa?]
cicerella (Rondani, 1876) (Gelechia) Italy
dubiosella (Beutenmüller, 1889) (Cerostoma) USA (Ak)
albovenosa Walsingham, 1907 USA (Hi)
dudiosella (Moriuti, 1977) (Cerostoma), missp.

THALASSONYMPHA Meyrick, 1931

19 **mysteriodes** Meyrick, 1931 Argentina

YPSOLOPHA Latreille, 1796
Ypsolophus Fabricius, 1798
Cerostoma Latreille, 1802
Hypsolopha Billberg, 1820
Theristis Hübner, [1825]
Harpipterix Hübner, [1825]
Abebaea Hübner, [1825]
Harpipteryx Treitschke, 1833, emend.
Chaetochilus Stephens, 1834
Harpepteryx Sodoffsky, 1837, emend.
Hypolepia Guenée, 1845, nom. nud.
Pteroxia Guenée, 1845, nom. nud.
Harpopteryx Agassiz, 1846, emend.
Hypsilophus Agassiz, 1846, emend.
Periclymenobius Wallengren, 1880
Credemnon Wallengren, 1880
Trachoma Wallengren, 1880
Pluteloptera Chambers, 1880
Plutelloptera Walsingham, 1881, missp.
Alapa Kieffer & Jörgensen, 1910
Mapa Strand, 1911, redesc. [for Alapa]
Pycnopogon Chrétien, 1922
Credemna Forbes, 1923, missp.
Melitonympha Meyrick, 1927
Chalconympha Meyrick, 1931
Credemon Moriuti, 1977, missp.

20 **cordillerella** (Kieffer & Jörgensen, 1910) (Alapa) Argentina
cordillerella (Strand, 1911) (Mapa), redesc. Argentina
21 **crispulella** (Berg, 1875) (Cerostoma) Argentina
22 **eurypepla** (Meyrick, 1931) (Chalconympha) Argentina
23 **hydraea** (Meyrick, 1919) (Cerostoma) Peru
24 **malacodoxa** (Meyrick, 1932) (Cerostoma) Argentina
25 **telluris** (Clarke, 1965) (Melitonympha), **n. comb.** Chile (Juan Fernandez Is.)

Praydinae

ATEMELIA Herrich-Schäffer, 1853

26 **contrariella** Zeller, 1877 Colombia

PRAYS Hübner, [1825]

27 **stratella** Zeller, 1877 Colombia

36. YPONOMEUTIDAE

by J. B. Heppner

Attevinae[36]

ATTEVA Walker, 1854
Poeciloptera Clemens, 1860, preocc. (Latreille, 1829 [Hemiptera])
Amblothridia Wallengren, 1861
Corinea Walker, 1863
Oeta Grote, 1865
Carthara Walker, 1866, preocc. (Walker, 1865 [Pyralidae])
Synadia Walker, 1866, repl. name
Scintilla Guenée, 1879, preocc. (Deshayes, 1856 [Mollusca])
Syblis Guenée, 1879

1 **cosmogona** Meyrick, 1931 Brazil (RJ)
2 **ergatica** Walsingham, 1914 Belize
3 **flavivitta** (Walker, 1866) (Carthara) Colombia

4 **fulviguttata** (Zeller, 1873) (Oeta) Surinam
glaucopidella (Guenée, 1879) (Syblis) Jamaica
5 **hysginiella** (Wallengren, 1861) (Amblothridia) Panama
sylpharis (Butler, 1877) (Cydosia) Galapagos Is.
6 **microsticta** Walsingham, 1914 Mexico (Pue)
7 **numeratrix** Meyrick, 1930 Brazil (Pa)
8 **punctella** (Cramer, 1781) (Phalaena) Surinam
pastulella (Fabricius, 1787) (Tinea) Surinam
pustulella (Fabricius, 1794) (Tinea) ?
subtilis (Hübner, [1819]) (Crameria) Surinam
aurea (Fitch, 1856) (Deiopeia) USA (Ga)
compta (Clemens, 1860) (Poeciloptera) USA (Tx)
gemmata (Grote, 1873) (Oeta) Cuba
aurera (Stretch, 1873) (Oeta), missp.
fastuosa (Zeller, 1877) (Oeta) Cuba
compta var. floridana (Neumoegen, 1891) (Oeta) USA (Fl)
edithella Busck, 1908 USA (Tx)
exquisita Busck, 1912 Mexico (Coa)
9 **siderea** (Walsingham, 1892) (Oeta) Virgin Is. (St. Thomas)

LACTURA Walker, 1854
Dianasa Walker, 1854
Mieza Walker, 1854
Sarbena Walker, 1865, preocc. (Walker, 1862 [Noctuidae])
Themiscyra Walker, 1865
Cyptasia Walker, 1866
Buxeta Walker, 1866
Enaemia Zeller, 1872
Pseudotalara Druce, 1885
Pseudocaprima Walsingham, 1900
Epidictica Turner, 1903
Hedycharis Turner, 1903
Eriopyrrha Meyrick, 1913

10 **atrolinea** (Barnes & McDunnough, 1913) (Mieza) USA (Tx)
11 **basistriga** (Barnes & McDunnough, 1913) (Mieza) USA (Tx)
12 **chrysippa** (Druce, 1885) (Pseudotalara) Guatemala
13 **citrina** (Busck, 1913) (Mieza) Costa Rica
14 **euthoracica** (Schaus, 1912) (Pseudotalara) Costa Rica
15 **irrorata** (Busck, 1913) (Mieza) Costa Rica
16 **lateralis** (Dyar, 1912) (Pseudotalara) Mexico (Ver)
17 **pseudophile** (Dyar, 1912) (Pseudotalara) Mexico (Oax)
18 **rutila** (Bartlett-Calvert, 1893) (Hyponomeuta [sic]) Chile
19 **schausia** Busck, 1920 Guatemala
20 **schenoxantha** (Schaus, 1912) (Pseudotalara) Costa Rica
21 **spatula** (Busck, 1913) (Mieza) Costa Rica
22 **subfervens** (Walker, 1854) (Mieza) "N. Am."
subferreus (Herrich-Schäffer, 1866) (Mieza), missp.

PYGMOCRATES Meyrick, 1932

23 **lissopeda** Meyrick, 1932 Brazil (SC)

Yponomeutinae

ANCHIMACHETA Walsingham, 1914

24 **capnodes** Walsingham, 1914 Mexico (Gue)
25 **iodes** Walsingham, 1914 Mexico (Gue)
26 **tolmetes** Walsingham, 1914 Mexico (Gue)

DITRIGONOPHORA Walsingham, 1897

27 **marmoreipennis** Walsingham, 1897 Grenada

EUARNE Möschler, 1890

28 **obligatella** Möschler, 1890 Puerto Rico

ITHUTOMUS Butler, 1883
Ithytomus Meyrick, 1914, emend.

29 **formosus** Butler, 1883 Chile

SPILADARCHA Meyrick, 1913

30 **derelicta** Meyrick, 1913 Guyana

SYNCERASTIS Meyrick, 1931 [37]

31 **ptisanopa** Meyrick, 1931 Chile

TEINOPTILA Sauber, 1902

32 **calcarata** (Meyrick, 1924) (Hyponomeuta [sic]) Bermuda

TOECORHYCHIA Butler, 1883

33 **cinerea** Butler, 1883 Chile

URODUS Herrich-Schäffer, 1854 [38]
Trichostibas Zeller, 1863
Pexicnemidia Möschler, 1890, **n. syn.** [39]
Paratiquadra Walsingham, 1897

34 **amphilocha** Meyrick, 1924 Brazil (Pa)
35 **aphanoptis** Meyrick, 1930 Brazil (Pa)
36 **auchmera** Walsingham, 1914 Guatemala
37 **brachyanches** Meyrick, 1931 Brazil (RJ)
38 **calligera** (Zeller, 1877) (Trichostibas) Cuba
39 **carabopa** Meyrick, 1925 Peru
40 **chiquita** (Busck, 1910) (Trichostibas) Costa Rica
41 **chrysoconis** Meyrick, 1932 Peru
42 **costaricae** (Busck, 1910) (Trichostibas) Costa Rica
43 **cumulata** Walsingham, 1914 Mexico (Ver)
44 **cyanombra** (Meyrick, 1913) (Trichostibas) Argentina
45 **cyclopica** Meyrick, 1930 Brazil (Pa)
46 **decens** Meyrick, 1925 Costa Rica
47 **distincta** (Strand, 1911) (Trichostibas) Panama
48 **favigera** (Meyrick, 1913) (Trichostibas) Peru
49 **fonteboae** (Strand, 1911) (Trichostibas) Brazil (Am)
50 **forficulella** (Walsingham, 1897) (Paratiquadra) Jamaica
51 **fulminalis** Meyrick, 1931 Brazil (MG)
52 **fumosa** (Zeller, 1863) (Trichostibas) Venezuela
53 **hephaestiella** (Zeller, 1877) (Trichostibas) Panama
54 **hexacentris** Meyrick, 1931 Brazil (MG)
55 **hypsicrates** Meyrick, 1925 Colombia
56 **imitans** (R. Felder & Rogenhofer, 1875) (Trichostibas) Colombia
57 **imitata** Druce, 1884 Guatemala
58 **iophlebia** (Zeller, 1877) (Trichostibas) "Antilles"
59 **isoxesta** Meyrick, 1932 Bolivia/Costa Rica
60 **isthmiella** (Busck, 1910) (Trichostibas) Panama
61 **lithophaea** (Meyrick, 1913) (Trichostibas) Guyana
62 **marantica** Walsingham, 1914 Panama
63 **merida** (Strand, 1911) (Trichostibas) Venezuela
64 **mirella** (Möschler, 1890) (Pexicnemidia), **n. comb.** Puerto Rico
65 **modesta** Druce, 1884 Guatemala
66 **monura** Herrich-Schäffer, 1854 Venezuela
67 **niphatma** Meyrick, 1925 Colombia
68 **opticosema** Meyrick, 1930 Brazil (Pa)
69 **ovata** (Zeller, 1877) (Trichostibas) Cuba
70 **pallidicostella** (Walsingham, 1897) (Trichostibas) Jamaica
71 **perischias** Meyrick, 1925 Brazil (Am)
72 **porphyrina** Meyrick, 1932 Colombia/Costa Rica
73 **praetextata** (Meyrick, 1913) (Trichostibas) Peru
74 **pulvinata** Meyrick, 1924 Peru
75 **sanctipaulensis** (Strand, 1911) (Trichostibas) Brazil (Am)
76 **scythrochalca** Meyrick, 1932 Costa Rica
77 **sordidata** (Zeller, 1877) (Trichostibas) Puerto Rico
78 **spumescens** Meyrick, 1925 Peru
79 **staphylina** Meyrick, 1932 Brazil (SC)
80 **sympiestis** Meyrick, 1925 Brazil (Am)
81 **transverseguttata** (Zeller, 1877) (Trichostibas) Panama
transversiguttata Walsingham, 1914, emend.
82 **triancycla** Meyrick, 1931 Paraguay
83 **venatella** (Busck, 1910) (Trichostibas) Brazil (Pr)
84 **xiphura** Meyrick, 1931 Brazil (SC)

XYROSARIS Meyrick, 1907
Xyrosaria Kearfott, [1903], missp.

85 **mnesicentra** Meyrick, 1913 Guyana

YPONOMEUTA Latreille, 1796
Hyphantes Hübner, [1806], suppr. (ICZN Op. 97)
Erminea Haworth, [1811]
Hyponomeuta Billberg, 1820, emend.
Coenyphantes Hübner, [1822]
Nygmia Hübner, [1825], preocc. (Hübner, [1820] [Lymantriidae])
Hyponomeuta Sodoffsky, 1837, emend.
Hyponomenta Turner, 1898, missp.

86 **eusoma** Walsingham, 1914 Mexico (Ver)
878 **triangularis** Möschler, 1890 Puerto Rico

ZELLERIA Stainton, 1849

88 **cirrhoscia** Meyrick, 1931 Argentina
89 **leucoschista** Meyrick, 1931 Argentina
90 **pistopis** Meyrick, 1931 Argentina

37. ARGYRESTHIIDAE

by J. B. Heppner

ARGYRESTHIA Hübner, [1825] [40]
Argyrosetia Stephens, 1829
Oligos Treitschke, 1830
Ederesa Curtis, 1833
Ismene Stephens, 1834
Blastotere Ratzeburg, 1840
Argyrestia MacKay, 1972, missp.

1 **biruptella** Zeller, 1877 Colombia
2 **carcinomatella** Zeller, 1877 Colombia
3 **conspersa** Butler, 1883 Chile
4 **diffractella** Zeller, 1877 Colombia
5 **melitaula** Meyrick, 1918 Colombia

6 **ochridorsis** Zeller, 1877 Colombia
7 **percussella** Zeller, 1877 Colombia

38. DOUGLASIIDAE

by J. B. Heppner

PROTONYCTIA Meyrick, 1931

1 **originalis** Meyrick, 1931 Ecuador

39. ACROLEPIIDAE

by J. B. Heppner

ACROLEPIOPSIS Gaedike, 1970
Argiope Chambers, 1873, preocc. (Audouin, 1827 [Arachnida])

1 **aureella** (Blanchard, 1852) (Elachista), **n. comb.** Chile
2 **bythodes** (Meyrick, 1919) (Acrolepia), **n. comb.** Peru
3 **cestrella** (Busck, [1934]) (Acrolepia), **n. comb.** Cuba
4 **chalcolampra** (Meyrick, 1931) (Acrolepia), **n. comb.** Chile
5 **chariphanes** (Meyrick, 1931) (Acrolepia), **n. comb.** Chile
6 **elaphrodes** (Meyrick, 1919) (Acrolepia), **n. comb.** Peru
7 **halosema** (Meyrick, 1931) (Acrolepia), **n. comb.** Argentina
8 **jaspidata** (Meyrick, 1919) (Acrolepia), **n. comb.** Peru
9 **maculella** (Blanchard, 1852) (Elachista), **n. comb.** Chile
10 **marmaropis** (Meyrick, 1919) (Acrolepia), **n. comb.** Colombia
11 **mixotypa** (Meyrick, 1931) (Acrolepia), **n. comb.** Chile
12 **niphosperma** (Meyrick, 1931) (Acrolepia), **n. comb.** Argentina
13 **poliopis** (Meyrick, 1919) (Acrolepia), **n. comb.** Peru
14 **prasinaula** (Meyrick, 1927) (Acrolepia), **n. comb.** Colombia
15 **seraphica** (Meyrick, 1931) (Acrolepia), **n. comb.** Argentina
16 **syrphacopis** (Meyrick, 1919) (Acrolepia), **n. comb.** Peru
17 **tharsalea** (Walsingham, 1914) (Acrolepia), **n. comb.** Guatemala
18 **xiphias** (Meyrick, 1931) (Acrolepia), **n. comb.** Chile

ANTISPASTIS Meyrick, 1926

19 **clarkei** Pastrana, 1952 Argentina
20 **selectella** (Walker, 1863) (Adela) [41] Brazil (Am)
21 **xylophragma** Meyrick, 1926 Peru

40. HELIODINIDAE

by J. B. Heppner

Schreckensteiniinae [42]

SCHRECKENSTEINIA Hübner, [1825]
Chrysocorys Curtis, 1833

1 **inferiorella** Zeller, 1877 Colombia
2 **jocularis** Walsingham, 1914 Guatemala

Heliodininae

AMPHICLADA Meyrick, 1912

3 **fervescens** Meyrick, 1912 Grenada

COPOCENTRA Meyrick, 1909

4 **calliscelis** Meyrick, 1909 Bolivia
5 **notopyrsa** Meyrick, 1936 Peru
6 **porphyropis** Meyrick, 1922 Brazil (Am)
7 **saltatoria** Meyrick, 1922 Brazil (Pa)
8 **submetallica** Meyrick, 1922 Peru

CREMBALASTIS Meyrick, 1915

9 **erythrorma** Meyrick, 1915 Peru

CYCLOPLASIS Clemens, 1864

10 **basiplagata** Walsingham, 1897 Virgin Is. (St. Thomas)
basiplagiata Forbes, 1930, missp.
11 **gnathodes** Meyrick, 1917 Ecuador
12 **habrarcha** Meyrick, 1917 Ecuador
Ecuador

HELIODINES Stainton, 1854 [43]
Aetole Chambers, 1875
Aetola Frey, 1884, missp.
Heliodinides Turner, 1941, missp.

*13 **albaciliella** Busck, 1910 USA (NM)
albiciliella Meyrick, 1913, emend.
14 **aureoflamma** Walsingham, 1897 Virgin Is. (St. Thomas)
*15 **bella** (Chambers, 1875) (Aetole) USA (Tx)
16 **choneuta** Meyrick, 1915 Colombia
*17 **ciccella** Barnes & Busck, 1920 USA (Az)
18 **demarcha** Meyrick, 1917 Peru
19 **isoleura** Meyrick, 1917 Ecuador
20 **loriculata** Meyrick, 1932 Bolivia
*21 **metallicella** Busck, 1909 USA (Az)
*22 **perichalca** Meyrick, 1922 USA (NM)
23 **quinqueguttata** Walsingham, 1897 Jamaica
23.1 **rubella** (Blanchard, 1852) (Elachista), **n. comb.** [44] Chile
24 **schulzella** (Fabricius, 1794) (Tinea) "West Indies"
*25 **sexpunctella** Walsingham, 1892 USA (Az)
*26 **tripunctella** Walsingham, 1892 USA (Tx)
27 **urichi** Busck, 1910 Trinidad

LAMPROLOPHUS Busck, 1900
Embola Walsingham, 1909

28 **dentifera** (Walsingham, 1909) (Embola) Mexico (Mor)
29 **marginata** (Walsingham, 1892) (Heliodines) St. Vincent
30 **obolarcha** Meyrick, 1909 Bolivia
31 **xanthocephala** (Walsingham, 1909) (Embola) Mexico (Tab)

LITHARIAPTERYX Chambers, 1876

*32 **abroniaeella** Chambers, 1876 USA (Co)
abroniella Meyrick, 1913, emend.
*33 **jubarella** Comstock, 1940 USA (Ca)
*34 **mirabilinella** Comstock, 1940 USA (Ca)

PSEUDASTASIA Walsingham, 1909

35 **opulenta** Walsingham, 1909 Panama

SCELORTHUS Busck, 1900

36 **calcifera** Walsingham, 1909 Mexico (Mor)
*37 **pisoniella** Busck, 1900 USA (Fl)

THRASYDOXA Meyrick, 1912

38 **tyrocopa** Meyrick, 1912 Colombia

Immoidea

41. IMMIDAE

by J. B. Heppner

MOCA Walker, 1863
Adricara Walker, 1863
Alicadra Walker, [1866]
Jobula Walker, 1866
Callartona Hampson, [1893]

1 **albodiscata** (Walker, 1863) (Adricara) Brazil (Am)
2 **antiquata** (Meyrick, 1913) (Imma) Guyana
3 **aphrodora** (Meyrick, 1922) (Imma) Brazil (Pa)
4 **ethirastis** (Meyrick, 1922) (Imma) Peru
5 **mitrodeta** (Meyrick, 1922) (Imma) Peru
6 **mniograpta** (Meyrick, 1931) (Imma) Peru
7 **nephallactis** (Meyrick, 1906) (Imma) Venezuela
8 **nipharcha** (Meyrick, 1931) (Imma) Brazil (Am)
9 **niphostoma** (Meyrick, 1922) (Imma) Brazil (Am)
10 **paratma** (Meyrick, 1912) (Imma) Guyana
11 **pelomacta** (Meyrick, 1922) (Imma) Brazil (Pa)
12 **phthorosema** (Meyrick, 1912) (Imma) Colombia
13 **roscida** (Meyrick, 1922) (Imma) Brazil (Am)
14 **rugosella** (Busck, 1913) (Imma) Guyana
15 **vexatalis** (Walker, [1866]) (Alicadra) Brazil
16 **zophodes** (Meyrick, 1909) (Imma) Bolivia

LOXOTROCHIS Meyrick, 1906

17 **sepias** Meyrick, 1906 Brazil (ES)

IMMA Walker, [1859]
Pingrassa Walker, [1859]
Tortricomorpha C. Felder, 1861
Topaza Walker, 1864
Vinzela Walker, [1866]
Thylacopleura Meyrick, 1886
Davendra Moore, 1887
Pseudotortrix Turner, 1900

18 **atialis** (Walker, 1859) (Pyralis) Brazil
19 **boeta** (Druce, 1898) (Thalpochares) Panama
20 **cancanopis** Meyrick, 1906 Colombia
21 **chloromelalis** (Walker, [1866]) (Aglossa) Brazil (Pa)
22 **cincta** (Druce, 1898) (Eustrotia) Guatemala
23 **ciniata** (Druce, 1898) (Thalpochares) Panama
24 **confluens** Meyrick, 1931 Brazil (Pa)
25 **cuneata** Meyrick, 1906 Brazil
26 **cyanospora** Meyrick, 1926 Colombia
27 **eriospila** Meyrick, 1922 Brazil (Pa)
28 **euglypta** Meyrick, 1931 Colombia
29 **leniflua** Meyrick, 1931 Colombia
30 **metachlora** Meyrick, 1906 Brazil
31 **prasinospora** Meyrick, 1915 Ecuador
32 **protocrossa** Meyrick, 1909 Bolivia

33 **quadrivittana** (Walker, 1863) (Gauris) Brazil (RJ)
34 **sciophanes** Walsingham, 1914 Guatemala
35 **thymora** Meyrick, 1906 Brazil (Am)
36 **varipes** (Walker, 1862) (Pyralis) Brazil

Addendum

22. ELACHISTIDAE

ARISTOPTILA Meyrick, 1932

0.1 **smaragdophanes** Meyrick, 1932 Ecuador

ATMOZOSTIS Meyrick, 1932

0.2 **hilda** Meyrick, 1932 Colombia

ELACHISTA Treitschke, 1833

3.1 **petalistis** Meyrick, 1932 Guyana

23. BLASTOBASIDAE

AUXIMOBASIS Walsingham, 1892

53.1 **invigorata** Meyrick, 1932 Virgin Is. (St. Thomas)

BLASTOBASIS Zeller, 1855

63.1 **athymopa** Meyrick, 1932 Virgin Is. (St. Thomas)

26. AGONOXENIDAE

DIACHOLOTIS Meyrick, 1937

0.1 **iopyrrha** Meyrick, 1937 Costa Rica

HELCANTHICA Meyrick, 1932

0.2 **spermotoca** Meyrick, 1932 Virgin Is. (St. Thomas)

PALAEOMYSTIS Meyrick, 1931

9.1 **chalcopeda** Meyrick, 1931 Brazil (RJ)

PROCHOLA Meyrick, 1915

22.1 **chalcothorax** Meyrick, 1932 Brazil (ES)
34.1 **sancticola** Meyrick, 1932 Virgin Is. (St. John)

ZARATHA Walker, 1864

39.1 **cervinella** (Walsingham, 1897) (Syntomactis), **n. comb.** Virgin Is. (St. Croix)

27. COSMOPTERIGIDAE

COSMOPTERIX Hübner, [1825]

6.1 **albicaudis** Meyrick, 1932 (Cosmopteryx [sic]) Virgin Is. (St. Thomas)

29. GELECHIIDAE

RECURVARIA Haworth, 1828

204.1 **hippurista** Meyrick, 1932 Virgin Is. (St. Thomas)

UNTOMIA Busck, 1906

494.1 **formularis** Meyrick, 1929 Brazil (Am)/Colombia

HYPATIMA Hübner, [1825]

503.1 **disposita** (Meyrick, 1931) (Chelaria), **n. comb.** Brazil (ES)

MYROPHILA Meyrick, 1923

617.1 **caryoplecta** Meyrick, 1930 Brazil (Pa)

NOTES

1. "Dalaca" species of authors that are not placed under **Dalaca** Walker, 1856, belong to other genera presently undefined, or the species need further study. [p. 17]
2. This is the first time Palaeosetidae have been represented in the Neotropical fauna (Nielsen and Robinson, 1983). [p. 17]
3. Nepticulinae is used in place of Stigmellinae, as proposed by Scoble (1983), due to the need for nominate subordinate taxa (ICZN Art. 37). [p. 17]
4. Nepticulini is used, as in footnote 3, in place of Stigmellini, as proposed by Scoble (1983) (ICZN Art. 37). [p. 17]
5. **Prodoxoides** is included, although described in 1984, due to the significance of including the first discovery of Prodoxidae from temperate South America. [p. 18]
6. Tineidae genera are listed alphabetically pending a generic revision, since the author considers the present subfamily limits too uncertain and ill defined. [p. 19]
7. The named subspecies of **Acrolophus macrogaster** (Walsingham) are listed as synonyms since they are not geographically isolated populations but apparently only forms of the same species. [p. 19]
8. **Choropleca terpsichorella** (Busck) was described from Hawaii from specimens introduced from Central America. [p. 21]
9. Many species remaining in **Tinea** Linnaeus may be found to belong to other genera following further study. [p. 23]
10. **Tinea galeatella** Mabille, 1888, has been listed as a synonym of **Plutella xylostella** (Linnaeus, 1758) but has been found to be a synonym of **Tinea pallescentella** Stainton, 1851 (teste E. S. Nielsen). [p. 23]
11. **Harmaclona** is retained in Arrhenophanidae pending further study to determine its possible placement in Tineidae. [p. 25]
12. **Dasycarea** is considered a valid representative of the Old World family Amphitheridae, the only known representative of this family in the New World. [p. 25]
13. The Madinier name, "noctuella," was originally used in the sense of "a small moth," and not as a binomial name, as subsequently cited. Article 11 (ICZN) indicates that the name is unavailable and, in not being binomial or scientific, a vernacular name. [p. 25]
14. **Cremastobombycia lantanella** Busck was described from Hawaii from specimens introduced from Mexico. [p. 26]
15. The placement of "Elachista" **rubella** Blanchard is still uncertain. It was thought to belong to **Phyllocnistis** but now appears best placed in Heliodinidae, tentatively in **Heliodines** (teste D. R. Davis). [p. 27]
16. Most Neotropical Oecophoridae generic type species have been examined by the author for assignment to subfamily. [p. 27]
17. **Exaeretia** Stainton, 1849, is considered by the author to be the senior synonym of **Depressariodes** Turati, 1924, and **Martyrhilda** Clarke, 1941, but some authors restrict **Exaeretia** to a few Palearctic species only. [p. 27]
18. Oecophoridae genera described by Clarke (1978) were not assigned to subfamilies by Clarke. They are placed in the checklist according to their apparently valid subfamily and tribal assignments but in alphabetical order under each higher category. [p. 27]
19. **Eumimographe** Dognin, 1905, is considered a separate genus by some researchers. [p. 28]
20. **Atteria lydia** Druce was transferred from Tortricidae to Oecophoridae by Obraztsov (1966. Proc. US NM., 118:620). The species is herein assigned to the **Eumimographe** group of species, now in **Hypercallia.** [p. 28]
21. **Erysiptila** Meyrick was not included in the New World monograph of Ethmiidae (Powell, 1973) but is closely related to **Ethmia** (teste V. O. Becker). [p. 29]
22. **Macrocirca** Meyrick is another genus not treated by Powell (1973), but apparently is closely related to **Ethmia;** this genus may even be a synonym of **Ethmia** (teste V. O. Becker). [p. 30]
23. The new name, **Rectiostoma** Becker (1982. J. N. Y. Ent. Soc., 89:270), was named to replace the preoccupied **Setiostoma** Zeller, 1875. The Zeller description inadvertently appeared after the publication of Felder and Rogenhofer (1875. "Reise Novara...") wherein the name "Setiostoma" was used for species now assigned to Glyphipterigidae. [p. 35]
24. Numerous species of **Stenoma** will in the future be assigned to new genera but are retained in **Stenoma** pending further study. [p. 35]
25. The generic assignment of **Dasycera auricollis** Walsingham is uncertain and is based on Walsingham's congeneric inclusion of the species with "Dasycera" **newmanella** Clemens, now assigned to **Mathildana.** [p. 40]
26. **Eupolella** Fletcher appears to be related to **Glyphidocera** and is, thus, transferred to Symmocinae (Blastobasidae) (teste V. O. Becker). [p. 41]
27. Syntypes of **Coleophora picticornis** Walsingham are from Haiti, St. Thomas, and Grenada. [p. 42]
28. The Neotropical genera of Gelechiidae are arranged by subfamily except for a large group of genera too little known to make any reasonable assessment of their subfamily position; these latter genera are arranged alphabetically as unplaced genera. [p. 44]
29. **Crambodoxa** Meyrick may belong to Oecophoridae (teste V. O. Becker). [p. 52]

30. **Iphimachaera** Meyrick may be related to **Tinaegeria** in Oecophoridae (teste V. O. Becker). [p. 52]
31. **Leistogenes** Meyrick, 1927, may be a junior synonym of **Logisis** Walsingham, 1909 (teste V. O. Becker). [p. 52]
32. **Ptilostonychia** Walsingham may be related to **Glyphidocera**, in which case it would belong in Symmocinae (Blastobasidae) (teste V. O. Becker). [p. 53]
33. "Gaea" **lilloi** Köhler is transferred from Sesiidae to Gelechiidae (teste T. D. Eichlin) but is as yet unassigned to a gelechiid genus. [p. 53]
34. The superfamily name Copromorphoidea is used, since higher category names need not be based on priority of authorship. Thus, Carposinoidea, Alucitoidea and Glyphipterygoidea are not used. [p. 53]
35. **Sericostola** Meyrick is here transferred from Yponomeutidae to Glyphipterigidae following examination of the type-species (teste J. B. Heppner). [p. 54]
36. Attevinae, **new status**, is here assigned as a new subfamily of Yponomeutidae for **Atteva**, **Lactura**, **Pygmocrates**, and the Old World **Anticrates**. These genera all differ from other yponomeutids in possessing chaetosemata on the head vertex. [p. 55]
37. **Syncerastis** Meyrick is here transferred from Tineidae to Yponomeutidae following examination of the type-species (teste D. R. Davis). [p. 56]
38. The subfamily assignment of **Urodus** requires further study. [p. 56]
39. **Pexicnemidia** Möschler, 1890, is transferred from Tineidae to synonymy of **Urodus** Herrich-Schäffer, 1854, following examination of the type-species (teste D. R. Davis). [p. 56]
40. There are undoubtedly a number of Nearctic species of **Argyresthia** that also occur in the oak zone of northern Mexico. [p. 56]
41. **Adela selectella** Walker was transferred from Adelidae to Acrolepiidae as a species of **Antispastis** by Davis (1980. J. Lepid. Soc., 32:187). [p. 57]
42. Schreckensteiniinae is retained as a subfamily of Heliodinidae. Having spines on the hind tibiae does not in itself signify a separate family of Lepidoptera, especially while other characters show relationships to other Heliodinidae. The apparent similarities to Epermeniidae are unconfirmed. [p. 57]
43. New World **Heliodines** require study to determine if they are actually congeneric with the European type-species of the genus. [p. 57]
44. "Elachista" **rubella** Blanchard, as noted in note 15, is only tentatively assigned to **Heliodines** (teste D. R. Davis). [p. 57]
45. **Mnesichara** Walsingham, 1912, is now thought to be a distinct genus in Oecophorinae, including the species **M. dyctiota** Walsingham, 1912, and **M. ithymeta** (Meyrick, 1926). [p. 28]

INDEX TO SPECIES

Addendum* [p. 58]

INDEX TO GENERA

Addendum